# A DIRECT CONTACT APPROACH TO PHYSICS

Albert W. McKinney III
2017 June 16

**Library of Congress Cataloging-in-Publication Data**
McKinney, Albert William III (1929-)
  A Direct Contact Approach to Physics
  Library of Congress Control Number: 2017910137
  CreateSpace Independent Publishing Platform,
    North Charleston, South Carolina
  ISBN: 978-1548329013

# DEDICATION

This book is dedicated to Phyllis McKinney, my wife of 66 years.

# ACKNOWLEDGMENT

The author extends his thanks to Dr. Jeremy Dunning-Davies for many helpful remarks provided before this book was begun.

# TABLE OF CONTENTS

# TABLE OF CONTENTS

Page    Topic

**TABLE OF CONTENTS**

**FIGURES**

**TABLES**

# TABLE OF CONTENTS

## PREFACE

This book asserts that communication between particles in the universe takes place not by photons, but instead by direct contact. Thus:

It denies the existence of photons

It classifies elementary particles as follows:

> Four are primitive: electrons, protons, and their antiparticles
> All other particles are close combinations of primitive particles

The assertion leads to many interesting results, some of which are listed below:

It explains what causes the three basic forces of physics, and how they work:

> The strong nuclear force
> Electromagnetism
> Gravity

It shows that the universe is not expanding:

> How faulty measurements led to that idea.

> What is really happening:

> > Fritz Zwicky was right: light loses energy as it travels through space
> > Dark matter and dark energy are unnecessary concepts
> > Stars in remote galaxies do not move at extreme velocities
> > Hence the universe is not expanding

It shows the correct reason for the precession of the perihelia of objects orbiting the Sun:

> It is not caused by general relativity

> > General relativity is not exact, but is a useful approximation
> > The assumption that spacetime is curved is not correct

> The cause is that sufficient mass blocks gravity

> Hence the planets orbit around an apparent center of mass of the Sun, which is about 4,375 meters from the Sun's true center of mass

The theory is descriptive: It shows what is going on. It is not a broad-brush approximation for finding numerical values.

One significant result of the theory has not been included in this book. It is a model of the hydrogen atom, in which the proton and electron are held together by repeated connections. Thus, the particles move in straight lines between successive connections. Hence there is no radiation emitted as the particles change direction. Thus there is no need for quantum mechanics here.

Unfortunately, this subject did not seem to fit in the confines of this book. However, a treatment of this subject is found in a previous book of mine, *Gravity*.

### Feedback

This book has evolved over a forty-seven-year period. Inconsistencies may have arisen despite my efforts to eliminate them. And as a fellow mathematician once told me, typographical errors in a book seem to be discovered at a steady rate over the years.

In the 47 years of this book's evolution, I have published 10 previous books. These books do not always agree with this one. Any differences must be resolved in favor of the present work.

If you, the reader, come across any errors (typographical, formatting, or in the theory itself), please let me know via e-mail at mickeymck@prodigy.net. My hope is to respond to all such messages, but I warn you that my wife and I sometimes take long trips. Hence please do not despair if a response is not immediately forthcoming!

# INTRODUCTION

This is a book of conjectures. These conjectures lead to a model of physics. This model offers explanations of many phenomena in physics, enough so that it seems reasonable to assert that these conjectures are correct.

There are two principal conjectures. The first is that four particles are the building blocks of the universe; namely, electrons, protons, and their antiparticles. These particles will be called *primitive* particles.

The second principal conjecture is that every primitive particle emits a *probe* at a rate equal to its frequency. For nonadjacent particles, such probes are the means of communication between them.

For particles which are adjacent ("touching each other"), communication is by direct contact.

A connection is formed between two particles when one of these three things happens: (1) The probe of one particle comes very close to another primitive particle; (2) The probe of one particle comes very close to the probe of another particle; (3) For two adjacent particles, the probe of either particle returns from its cycle.

In each case, the connection is *very* brief, but lasts long enough for energy to be transferred from one particle to the other. And the three conservation laws (energy, linear momentum, and angular momentum) are imposed automatically and instantaneously when such a connection occurs.

Once a connection occurs, it may result in a gravitational interaction, or an electromagnetic interaction, or a strong nuclear force interaction. Or there may be no resulting force between the particles.

The theory behind these ideas is called *connection theory*.

Before going into details about these ideas, it is necessary to point out a couple of myths currently used in physics, and to offer a better myth to replace them.

1

## PROLOGUE: MYTHS IN PHYSICS

Underlying any scientific theory is a sct of ideas which cannot be proved or measured. They are the postulates on which the theory is based. Belief in these ideas is necessary in order to accept the theory. For better or worse, such belief is simply a matter of faith, due to the lack of any kind of proof or measurement.

According to the Wiktionary:

> http://en.wiktionary.org/wiki/Wiktionary:Main_Page,

one definition of *myth* is "A person or thing existing only in imagination, or whose actual existence is not verifiable."

Thus it seems fair to say that these ideas or postulates comprise a myth.

### Two Current Myths in Physics

### Myth 1: Communications

One myth on which modern physics is based holds that communication between particles is accomplished either by extremely small particles, such as photons, gravitons, neutrinos, antineutrinos, quarks, and gluons, or by fields. This myth is a foundation of the standard model of modern physics.

### Why is this a myth?

There are several reasons. Consider the photon. It travels from its source to its target at the speed of light. Between those two points, it is invisible. There is no way to detect its passage except by intercepting it. But intercepting it causes it to deposit its energy and then cease to exist. If it is not intercepted, then when it reaches its target, it deposits its energy and then ceases to exist.

Does the photon have mass? Apparently not, yet it has energy. How can a particle not have mass? What does it mean to say that a particle consists of pure energy?

Special relativity claims that the amount of time it takes for the photon to travel between its source and its target is zero in its own time frame. Yet to an observer, it takes $d/c$ seconds, where $c$ is the speed of light, and $d$ is the distance between the source and the target. It is not obvious that these different times make sense.

But taking this a step further, it seems to say that in its own time frame, the photon exists for exactly zero seconds. Again, it seems incredible to think that a particle can consist of pure energy and only exist for zero seconds.

The same considerations apply to gravitons and other similar particles.

3

The above ideas imply that there is no physical evidence proving that photons, gravitons, etc., exist in the form of particles. And of course, there is no physical evidence to the contrary.

### Myth 2: Extrapolations

It is traditional to consider that such things as forces and fields are independent entities, and to extrapolate their properties to cases of small distances.

For example, consider the force of gravity. It is well known that the gravitational force between two masses separated by a large distance varies with the inverse square of the distance. Extrapolating that to very small distances leads to the conclusion that gravitational force increases without limit as the masses approach each other.

Another force that is traditionally treated in a similar manner is electromagnetic force. Again, it is clear from experimental evidence that for two charged particles separated by a large distance, the electromagnetic force between them varies with the inverse square of the distance. Extrapolating that to the case of very small distances leads to the conclusion that electromagnetic force increases without limit as the two charged particles approach each other.

### Why is this a myth?

These forces have never been measured at near zero distances. But it has been assumed that such increases apply in the formation of black holes. It is supposed that when a star reaches some critical mass, it encounters an extreme gravitational force which causes it to shrink instantaneously to a point mass.

Of course, such an event has never been observed.

### A Better Myth for Physics

A different myth has been proposed as a basis for physics, one which replaces both of the above myths. This myth asserts that communication between particles takes place by means of the vibration of electrons, protons, and their antiparticles. This myth is the foundation of *connection theory*, which is the main subject of this volume.

Specifically, the assertion is that at each vibration, a particle emits a probe, and that the probe can make a connection with another particle. During such a connection, the connected particles generally exchange an impulse, and sometimes exchange the energy of a photon. Such exchanges are governed by the three laws of conservation (energy, linear momentum, angular momentum), which thus cause changes in the velocities and directions of the two particles.

### Why is this a myth?

This is a myth because nobody has ever measured the way in which particles vibrate, nor is it likely that anybody ever will. (Is this implied by the Heisenberg Uncertainty Principle?)

### Why is this myth of interest?

There are many reasons. For one, it leads to the conclusion that when two particles approach each other, the gravitational force between them tends to a finite limit.

Likewise, it leads to the conclusion that when an electron and a proton approach each other, the electrical force between them tends to a finite limit.

Furthermore, it offers a simple, straightforward definition of gravity. That definition carries with it a number of situations in which gravity is blocked. One such blockage accounts for the precession of the perihelion of every planet and asteroid that orbits the Sun.

Also, that definition precludes the possibility that gravity causes space to be curved. Thus although the theory of general relativity provides a rather accurate approximation to some effects of gravity, it is not an accurate picture of what gravity really is, nor how it works.

Another reason why this myth is of interest is that it yields a unified description of three physical forces, showing that gravity, electromagnetism, and the strong nuclear force all arise from the same physical mechanism.

Perhaps the strongest reason for accepting this myth is that it shows why the decay of an excited hydrogen atom to the ground state takes many billions of electron cycles, and why the decay of para-positronium takes about 15 billion electron cycles.

In the standard model of physics, it would seem that in each of those cases, the atom waits all that time in order to release a photon or two. What is it waiting for?

I believe that the correct answer is this: It is waiting to find a suitable recipient for the photon energy. Apparently it is not able to send out a random photon to seek for one. Instead, it sends out its probe billions of times, looking for a connection which will allow it to release the energy of its photon.

In connection theory, the idea of energy being transmitted by probes suggests that a suitable recipient is one which has a special relation to the probe.

Therefore, connection theory defines two types of connections, which will be called loose and tight. A loose connection does not allow for the transmission of energy such as that of a photon. A tight connection is one which does allow such a transmission.

The key to a tight connection may lie with the form of the probe tip, which must align with the target core in a very special way.

So seeking such a special connection may well explain why it takes billions of connections in the above examples.

# CHAPTER 1:  PHOTONS

The universe contains *matter*. (Matter is stuff made up of electrons and protons, for the most part.) Matter has various properties. Two of these are *mass* and *energy*. By Einstein's famous formula $E = mc^2$, energy $E$ is proportional to mass $m$. Thus, wherever there exists energy, there also exists mass.

The wave packet was introduced by Einstein in 1905 as the means by which energy is transferred from one particle to another. (It was later renamed photon.) Thus, a photon was presented as a mass-less bundle of energy. But by the previous paragraph, this is not possible. In particular, energy is a property, not a substance. It makes just as much sense to conceive of a length-photon (an object without mass which consists only of lengths).

So the idea of an energy photon is unreasonable.

Lacking photons, how is light (or any other form of energy) transferred from one particle to another? The only way that makes sense is by direct contact.

So how is energy transferred from one particle to another when the particles are not adjacent?

Well, consider that particles such as electrons and protons have *frequencies*. They vibrate!

What do these vibrations consist of?

It seems reasonable to consider that such a particle consists of two parts, which I call the core and the probe. The core is the center of mass of the particle, while the probe is an object which oscillates about the core. On its outward journey from the core, the probe may come in contact with another core, or with another probe. Such a contact allows for the exchange of energy between the probe and that other core or probe.

Thus the probe can carry energy from its core to another particle, and transfers it by direct contact.

Although photons do not exist, the word *photon* will be used in the remainder of this book to denote the energy transferred in such a contact.

# CHAPTER 2: CONNECTION THEORY

## Particles and Radiation

The universe contains particles and radiation. As it happens, every known example of radiation originates in a particle, and its effect becomes apparent when it is deposited on another particle. Thus every kind of radiation amounts to the transfer of energy from one particle to another. Hence it is convenient to begin the story of the universe with a consideration of particles.

## Particles

There are several different types of particles, some having properties that will be described below. Among these properties are ones that enable particles to communicate with one another. Such communications lead to interactions which are governed by the three basic conservation laws of physics, namely, the conservation of mass, linear momentum, and angular momentum.

### Primitive Particles

To begin with, there are four types of particles that turn out to be the building blocks of matter. These four types are electrons, protons, and their antiparticles, and will be called *primitive particles*. All other particles are combinations of primitive particles. A few details on such combinations must be postponed to Chapter 5.

### Two Properties of Primitive Particles

A primitive particle has two important properties. First, it has a *sign*: it is either *positive* or *negative*. Thus electrons and antiprotons are negative, while protons and positrons are positive. This property comes into play when particles communicate with one another. [Note: The choice of words here is arbitrary. The relevant property is really that there are two *kinds* of primitive particles. The words *positive* and *negative* are simply traditional names for these two kinds, but those two words are arbitrary; it would be just as informative to call the two kinds apples and oranges.]

Second, a primitive particle *vibrates*. The number of vibrations per second is called the frequency of the particle. This is calculated from the formula $E = mc^2 = h\nu$, where $E$ is the energy of the particle, $m$ is its mass, $c$ is the speed of light, $h$ is Planck's constant, and $\nu$ is the frequency. For example, an electron at rest has a frequency of $1.236 \times 10^{20}$ vibrations per second, while a proton at rest has a frequency of $2.269 \times 10^{23}$ vibrations per second. An antiparticle has the same frequency as its corresponding particle. A primitive particle in motion has a higher frequency, which is calculated by using the above formula with $m$ replaced by $\dfrac{m}{\sqrt{1-v^2}}$, where $v$ is the velocity of the particle.

What is the nature of these vibrations? To describe them, it is convenient to suppose that a primitive particle consists of two parts, called the *core* (or place holder or center of mass), and the *probe*.

9

A single vibration of a primitive particle is called a *cycle*. In each cycle, the core of the primitive particle emits a probe. The probe moves out in some direction from the core until either (1) it runs out of time, or (2) it connects with another particle (a probe-core connection), or (3) it connects with another probe (a probe-probe connection). (A connection occurs when the tip of the probe passes close enough to another particle, or to the tip of another outgoing probe.) In the latter two cases, there is an exchange of energy. In cases (1) and (2), the probe then returns to its core, and another cycle begins. In case (3), both probes continue.

Probes are one of the means by which particles communicate with one another. Another means of particle communication is core-core contact; this is described in Chapter 5.

Probes are emitted in a plane which is perpendicular to the spin axis of the particle. Of course, an interaction between two particles often changes the directions of the two spin axes, so over the long run, probes are emitted in more or less random directions.

However, on two successive cycles, the probe from a primitive particle is emitted in roughly opposite directions. Since these directions are not exactly opposite, the probe directions gradually move around the particle, an effect known as particle spin. Figure 2-1 (which is not to scale) indicates the directions of eight successive probes from a single primitive particle:

**Figure 2-1**
**Successive Probe Directions**

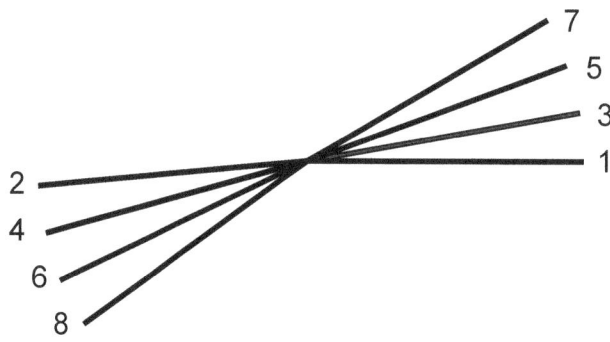

Since the cycle of an electron probe lasts for only a tiny fraction of a second, namely $1/(1.236 \times 10^{20}) = 8.09 \times 10^{-21}$ seconds, its time is up rather quickly! Even so, that is time enough for it to travel out and back about $1.87 \times 10^{26}$ meters, which is almost 20 billion light years.

The cycle of a proton probe lasts only $1/1836$ as long as that of an electron probe, and so can only go out a maximum of about $1.02 \times 10^{23}$ meters, or about 11 million light years.

Since a probe obeys the rule that nothing can exceed the speed of light in a vacuum, this means that it has to move forward in time: It goes into the future! Of course, that future is the average time as measured by the particles it passes on its journey. However, its own clock is only affected to the extent that it measures its own time. And when it returns to its core, its own time has passed to the extent that it equals the present time of its surroundings.

The above three paragraphs present the first three of several rather strange facts about physics!

### The Future

Every particle keeps its own time. When it sends out a probe, the probe goes out by its own clock, and does not go into its own future. However, its position as it travels puts it into a future relation with other particles. So there is no time travel involved here, but just a changing temporal relationship with other particles.

### What Does a Probe Consist of?

Physical evidence suggests that a probe consists of the mass of the particle. Thus when the probe is at its core, it consists of a lump of matter with mass equal to the measured mass of the particle. As it flies out from the core, it leaves behind an infinitesimal trail of mass, so that the lump which is the probe tip becomes less and less massive as the cycle continues. If a contact occurs at a distance $D$ from the core, the reaction involves a mass $m = m_e\left(1 - \dfrac{D}{M}\right)$ for the probe tip, where $m_e$ is the measured mass of the particle, and $M$ is the maximum distance that the probe can go. If the probe does go that maximum distance, then at that point its effective mass is zero. As it returns to its core, it accumulates the mass that it left behind on its outward journey.

### About the Core

Very little can be said at this point about the core. However, it is necessary to suppose that every particle core has a mass which is distinct from the measured mass of the particle. It seems likely that this mass is substantially smaller than the measured mass of the particle, which is identified with the mass of the probe.

### Random Events

For the cases that a probe connects with another particle or another probe, it is useful to have some names for the participants. The particle emitting the probe is called the source particle and its probe is called the source probe. A particle which connects with the source probe is called a target particle; a probe which connects with the source probe is called a target probe.

All interactions between particles are random events. For a connection to occur, the direction of the source probe must bring the probe tip to within some critical distance of the target core, or the tip of the target probe.

The probability of a probe-core connection depends on the distance $D$ between the two particles, and on the area $A$ of the source probe tip. The probability is equal to some constant $\kappa$ times $A$, divided by the surface area $4\pi D^2$ of a sphere of radius $D$. Thus, the probability is $\kappa A/4\pi D^2$. This is the reason for the inverse square property of many physical forces.

The probability of a probe-probe connection is more complicated, but essentially equals the probability that both probes pass very near a given point in space at the same time.

There are three possible connections between particles. Two of them were discussed in the previous two paragraphs: probe-core connections, which are the cause of electromagnetic effects, and probe-probe connections, which are the sole cause of gravity. The third is a core-core connection, which is responsible for the strong nuclear force.

A bit more about these three connections: A single probe-core connection results in a positive or negative electromagnetic interaction. Electromagnetic force results from the accumulation of such connections.

A single probe-probe connection yields an attractive impulse between each core and the point of the connection. It may result in an attractive or repulsive force between the two cores. However, the accumulation of such connections averages out to an attractive force between the two cores (gravity). This is described in detail in Chapter 6.

A single core-core connection typically occurs in a composite particle, where the cores of two or more primitive particles are present. In such a situation, when the probe of each of those particles returns to its core, it interacts with each of the other cores present. The result is the strong nuclear force.

### Forces at Extremely Small Separations

In contrast to the formulas of conventional physics, the connection-theoretic formulas for electromagnetic and gravitational force between two particles level off to finite values as the two particles approach each other; neither of these forces approaches infinity. The reason is that the probability of a connection increases as two particles approach each other, eventually reaching but never exceeding 0.5.

Why not 1? Well, even in the most extreme case, the target particle will be either on one side or the other side of the source particle. Hence, the target particle will be connected with the source particle at most on every other cycle of the source particle, and the probability of a connection on one cycle is thus never more than one-half.

Thus in general, the inverse square laws of physics apply only at distances which are greater than atomic distances.

This implies that the collapse of a large star does not lead to an infinitely small point of huge mass, but rather to an extremely condensed mass in which the particles are still individual.

### The Range of a Probe

This section analyzes redshift data obtained by Allan Sandage, which are found in Table 2-2 in the appendix to this chapter. These data are used to obtain an estimate of the maximum range of a probe. There are two maxima to be estimated, one for the electron probe, the other for the proton probe. These will be denoted by the symbols $M_e$ and $M_p$.

### Redshift Data

The redshift $z$ of incoming light is defined as

$$z = \frac{\lambda_{rec} - \lambda_{em}}{\lambda_{em}},$$

where $\lambda_{rec}$ is the received wavelength and $\lambda_{em}$ is the emitted wavelength. Recall that wavelength $\lambda$, frequency $v$, and energy $E$ are related by the familiar equations $c = \lambda v$ and $E = hv$, where $c$ is the velocity of light in a vacuum and $h$ is Planck's constant. With these equations, $\lambda$ can be transformed into units of energy as follows:

$$\lambda = \frac{c}{v} = \frac{ch}{E}.$$

Hence, the redshift $z$ can be expressed in terms of the energies:

$$z = \frac{\dfrac{ch}{E_{rec}} - \dfrac{ch}{E_{em}}}{\dfrac{ch}{E_{em}}} = \frac{E_{em}}{E_{rec}} - 1.$$

It is convenient to define the observed fraction $F$ of energy received:

$$F = \frac{E_{rec}}{E_{em}} = \frac{1}{1+z}.$$

The standard way of plotting redshift data is to put $\log z$ on the vertical axis, and stellar magnitude $V_C$ on the horizontal axis. See, for example, the papers by Allan Sandage in Volume 178 of the Astrophysical Journal (1972), pages 1-24.

An alternate way of representing such data is to plot stellar distance on the vertical axis, and the fraction of energy observed on the horizontal axis. Let $D$ denote distance in units of $10^{25}$ meters. Then $D$ is related to stellar magnitude $V_C$ by this equation:

(2-1) $\qquad D = 1.3031 \times 10^{\frac{V_C}{5}-3}$.

Using these formulas for $F$ and $D$, the data from Tables 2 through 4 (pages 6-10) of Sandage's paper have been converted to $F$ and $D$ values. The results are listed in Table 2-2 in the appendix to this chapter, and plotted in Figure 2-2.

Perhaps the most significant feature of Figure 2-2 is that the data appear to justify a linear fit of the form

(2-2) $\qquad D = \alpha(1-F)$

at least as well as they do the Hubble-Sandage curve, which for $D$ versus $F$ takes the form of a hyperbola:

$$D = \frac{\beta(1-F)}{F}.$$

A curve and a straight line have been plotted in Figure 2-2. The curve is the Hubble-Sandage curve, using the value $\beta = 16.819$ derived from Sandage's paper.

The line, labeled "Average $\alpha$", has a slope equal to the average of the $\alpha$ values in Table 2-2; this average value is 18.673. The reason for taking an average rather than, say, using the method of least squares, is this: If the objective were to establish a curve, then least squares would probably be used on the $D$ and $F$ data in order to derive the coefficient $\alpha$. However, least squares emphasizes the points at the ends of the scale over those in the middle. In this case, that means the points with the lowest and highest values of $F$. Of course, the points with low values of $F$ are perhaps the ones with the most uncertainty.

But the objective is not to establish a curve. Only the single value of $\alpha$ is needed. Therefore, instead of using least squares, a simply average of the $\alpha$ values for the 84 points seems sufficient. In this way, all points are given equal weight, regardless of their distance. Even so, this may be giving undue emphasis to the far points. However, most of them have $\alpha$ values that fall well within the range of the nearer clusters. In fact, the cluster which is farthest from the mean is the last one in the table, and it is just 2.9 standard deviations from the mean. (The standard deviation is 3.22.)

`FIGURE 2-2
Distance versus Energy Fraction

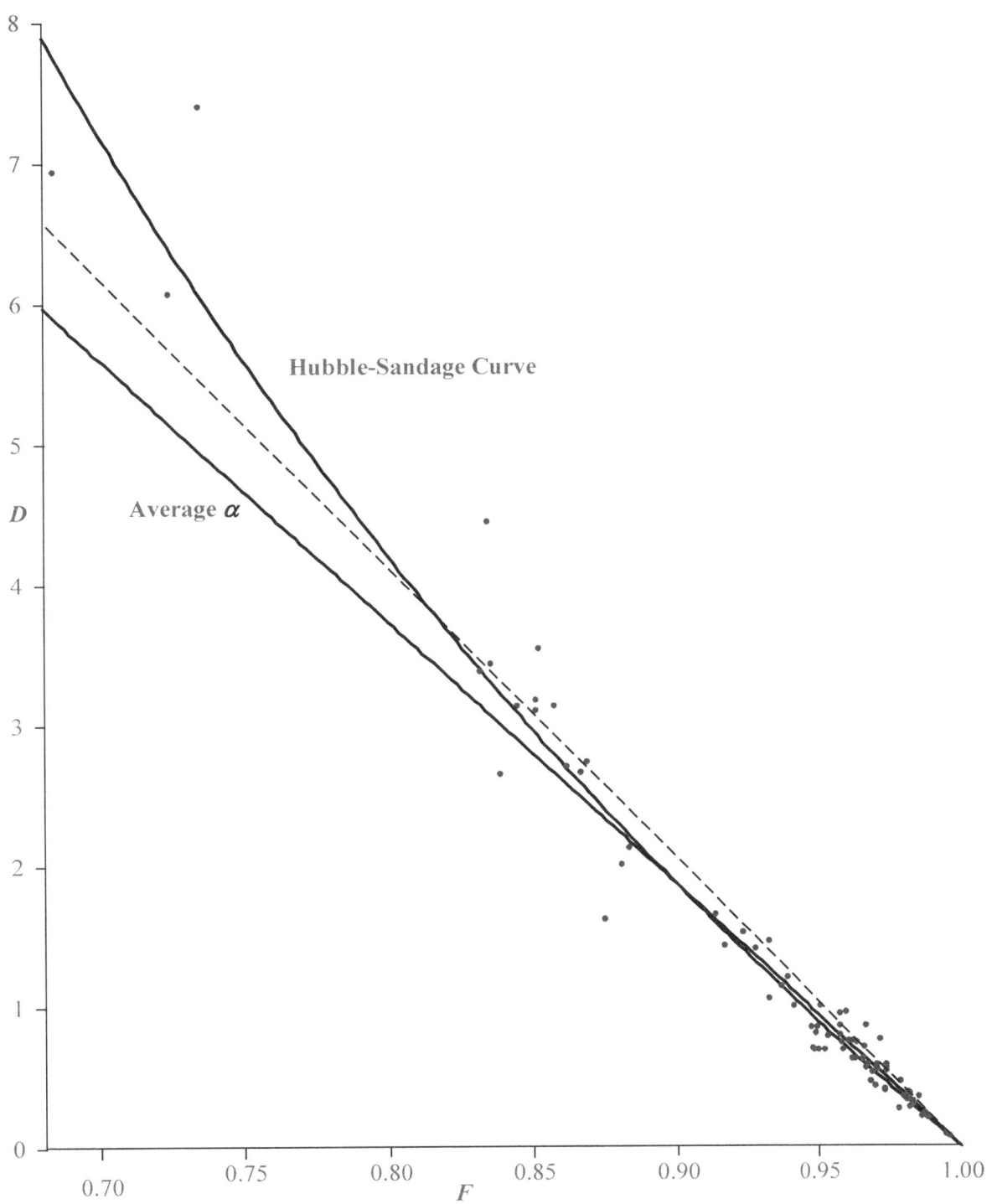

Extrapolating to $F = 0$ along the line corresponding to this average value of $\alpha$ yields a $D$ intercept of $1.8673 \times 10^{26}$ meters (19.74 billion light years). Thus, the estimated value for the range of an electron probe is $M_e = 1.8673 \times 10^{26}$ meters. One implication is that light cannot be transmitted directly over a distance longer than that. Also, equation (2-2) now can be written in the final form:

(2-3) $$D = M_e(1 - F) = M_e\left(1 - \frac{E_{rec}}{E_{em}}\right).$$

The dashed line in Figure 2-2 represents a test value which is 10% larger than $M_e$.

The next step is to use the redshift data to derive a possible mathematical formula for the motion of a probe tip. Equation (2-3) is taken as the starting point. A further assumption is also needed:

**A2.1**       The energy of a photon transmitted by a probe connection is proportional to the square of the velocity of the probe tip at the moment of impact.

The justification for this assumption is as follows. The velocity of a probe is equal to $c$ for most of its jorney. At *very* long distances, the velocity drops, and falls to zero by the end of the range. The above assumption is an arbitrary but simple rule that specifies that the energy varies as $v^2$, where $v = \dfrac{dD}{dt}$. The energy of a photon which at inception equals $h\nu$ thus ends up as

$$h\nu\left(\frac{v^2}{c^2}\right)$$

upon a connection. The energy fraction $F$ is thus

$$F = \frac{v^2}{c^2}.$$

Then by equation (2-3),

$$D = M_e\left(1 - \frac{v^2}{c^2}\right).$$

From this, it follows that

$$v = c\sqrt{1 - \frac{D}{M_e}},$$

and so

(2-4)                    $v \equiv \dfrac{dD}{dt} = c\sqrt{1 - \dfrac{D}{M_e}}$ .

Integrating this differential equation, and using the initial condition $D = 0$ when $t = 0$, yields

$$D = ct - \frac{c^2 t^2}{4M_e} \ .$$

Thus $D$ is a parabolic function of $t$. When $t = \dfrac{2M_e}{c}$, then $D = M_e$, and when $t = \dfrac{4M_e}{c}$, then $D = 0$. However, the last half of this parabola is not used, for the following reason: The variable $t$ is in units of ordinary time. As mentioned previously, the probe tip goes back in time as it returns to its core. Hence the use of $t$ for the second half of the parabola is not appropriate. However, when the probe tip arrives back at the core, a single cycle has elapsed. To model this behavior, the above equation is recast in terms of the proper time $\tau$ of the particle. This assumption is made about proper time:

**A2.2**      Proper time is proportional to wall clock time during the first half of the cycle.

That is, as the probe tip goes from the core to its maximum extent, its proper time goes through half of a cycle. Thus if $t = 0$ when $\tau = 0$, then $t = \dfrac{2M_e}{c}$ when $\tau = \dfrac{1}{2v}$, and so for the first half of the parabola, $\tau = \dfrac{ct}{4M_e v}$. However, $t = \dfrac{1}{v}$ when $\tau = \dfrac{1}{v}$, so the proportionality does not hold over the last half. Hence in proper time, the equation of the outward motion of the probe tip is

(2-5)                    $D = 4M_e v\tau(1 - v\tau)$ .

From this equation, the probe tip velocity at $\tau = 0$ can be found. The result is

$$\left.\frac{dD}{d\tau}\right|_{\tau = 0} = 4M_e v \ .$$

For an electron probe, this yields the velocity $4 \times 1.8673 \times 10^{26} \times 1.236 \times 10^{20} = 9.2319 \times 10^{46}$ meters per second. If this seems a bit large, remember that it is merely the speed of light, as measured in proper time.

### Particle Interactions Involving Probes

### Details of Particle Interactions

When a source probe connects with a target core, two things may happen: (1) the two particles *always* exchange energy; and (2) one particle *may* transfer additional energy to the other. The results of this interaction are governed by the laws of conservation of mass, linear momentum, and angular momentum. The energy exchange, and the energy transfer, occur in the form of impulses. The connection lasts only long enough for the interaction to take place. Following this, the source probe cannot engage in any more connections on that cycle, and soon returns to the source core.

Consider two particles which happen to interact in this way. If the two particles are stationary with respect to each other, then the resulting energy exchange involves two opposite impulses. These impulses are either both repulsive or both attractive, depending on whether the two particles have the same sign or not. If they have the same sign, then the resulting impulses tend to push the particles apart, whereas if they have opposite signs, then the impulses tend to pull them together. (Why are these two impulses opposite? Simply because one acts on one particle, and the other acts on the other particle, and by Newton's laws, these actions must be equal in force but opposite in direction.)

Note that the amount of energy exchanged has nothing to do with the concept of *charge*; it is solely determined by the above conservation laws.

[An implication of the above paragraph is that neither electrons nor protons nor their antiparticles have any charge—rather, they are electrically neutral! And this calls into question whether or not quarks exist, since they are presumed to have charges of $\pm \frac{1}{3}e$ or $\pm \frac{2}{3}e$. Clearly if a proton is electrically neutral, it cannot be composed of positively charged subparticles.]

Also, the amount of energy exchanged has nothing to do with the distance between the particles. This would seem to contradict common experience, which is that the farther apart the two particles are, the less effect the connection has. But the reason for less effect is only that the farther apart the two particles are, the less probable it is that there is any connection at all.

### Magnetism

In a probe-core connection, if the target core is moving with respect to the source core, then an additional phenomenon arises which causes four, not two, impulses to be involved.

The first two impulses are equal and opposite ones which act along the line between the two cores, just as in the case of a stationary target. These are called *electrical* impulses. The third and fourth impulses are equal and opposite ones which act along a line different from that between the two cores, and are called *magnetic* impulses. Figure 2-3 illustrates the situation:

**Figure 2-3**
**Impulses Resulting from a Probe-Core Connection**

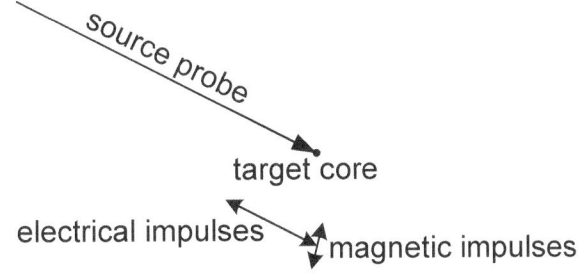

The upper and lower parts of this sketch should be superimposed, so that the intersection of the two lower arrows lies on the target core. The idea is that the electrical impulses act upon the two cores, whereas the magnetic impulses act on the target core and the source probe. The direction of the magnetic impulses depends on the direction of the target core with respect to the source core, by means of a somewhat complicated formula. Although the names differ, the electrical and magnetic impulses are physically simply impulses.

The above ideas can readily be developed to describe a magnetic field.

### Probe-Probe Interactions

There are two ways in which a probe can interact with another particle. One is as described above, where the source probe connects with a target core.

Another is where the source probe connects with a target probe. In this case, the two probes exchange attractive impulses, but neither of them returns immediately to its parent particle. Instead, both continue to travel, but in slightly different directions resulting from the impulse exchange. This phenomenon is responsible for gravity, and in particular, for the gravitational bending of light passing a star.

### Justification

The rules governing particles and probes in the above paragraphs of this chapter are not arbitrary. Each one has been formulated to reflect observations. The rules and their justification are presented in the following paragraphs.

### Rule: A probe-core connection causes the probe to end its cycle

This rule comes from an analysis of the deuteron (the nucleus of the deuterium atom). A deuteron consists of a proton plus a neutron. As explained in Chapter 5, connection theory holds that a neutron consists of a close combination of a proton and an electron. Thus the deuteron consists of a close combination of two protons and one electron.

A fundamental property of electrons and protons is that they oscillate: Each such particle emits probes at a frequency which is at a minimum when the particle is at rest, but is greater when the particle is moving. Therefore, it follows that the three particles of the deuteron must be emitting at least as many probes as they would if all three were at rest. Corresponding values are shown in Table 2-1.

**TABLE 2-1**
**Frequency: Number of Probes Emitted per Second by Particles at Rest**

| Particle | Frequency $\left(s^{-1}\right)$ |
|----------|--------------------------------|
| Electron | $1.235\,559 \times 10^{20}$ |
| Proton | $2.268\,732 \times 10^{23}$ |

This table shows that the three particles of the deuteron must emit at least

$$1.235\,559 \times 10^{20} + 2 \times 2.268\,732 \times 10^{23} = 4.538\,700 \times 10^{23} \equiv P_1$$

probes per second, which corresponds to a mass of $3.346\,155 \times 10^{-27}$ kg. (This follows from the special relativity formula $mc^2 = h\nu$, where $m$ is the particle mass, $c$ is the velocity of light, $h$ is Planck's constant, and $\nu$ is the particle frequency.) And it is quite likely that these nuclear constituents are moving, so that their frequencies are greater than the rest frequencies.

However, the measured mass of the deuteron is only $3.343\,583 \times 10^{-27}$ kg., which corresponds to $4.535\,211 \times 10^{23} \equiv P_2$ probes per second. Thus the deuteron would seem to have a deficiency of at least $P_1 - P_2 = 3.489 \times 10^{20}$ probes per second. How can this deficiency be explained?

To set the stage for an explanation, assume that the particles in the deuteron emit a total of $N_e$ probes per second, but that $N_a$ of them are absorbed by one of the other nuclear particles, and the remainder, $N_r = N_e - N_a$, miss the other particles. Assume further that the ones that miss are the only ones that can be observed, and that they account for the measured mass of the deuteron. In this case,

$$N_r = P_2 = 4.535\,211 \times 10^{23}$$
$$N_e \geq P_1 = 4.538\,700 \times 10^{23},$$

hence $N_a = N_e - 4.535\,211 \times 10^{23} \geq 3.489 \times 10^{20}$.

It seems reasonable to assume that the $N_a$ absorbed probes account for the binding energy of the deuteron, which is about 2.224 52 MeV. This is equivalent to $3.965\,566\,734 \times 10^{-30}$ kilograms, and to a frequency of $5.378\,865\,406 \times 10^{20}$ probes per second. That is, $N_a = 5.378\,865\,406 \times 10^{20}$.

This implies that sometimes a probe emitted by a nuclear particle will connect with another particle in the same nucleus and will not proceed outside of the nucleus. It was for this reason that it was stated earlier that when a probe connects with another particle, it stops there and does not continue on its outward journey. Of course, an alternative explanation is that if it does continue on that journey, it cannot engage in any further connections on that cycle.

A further implication is that the measured mass of an atom is based on the probes which emerge from the atom, but does not count the probes which connect with other particles in the nucleus. Therefore, the velocities of the particles in the atom are greater than would be suggested by the measured mass of the atom.

### Rule: A probe-probe connection does not limit the cycle of either probe

There are no physical events known to me which suggest that a probe-probe connection has any aftereffect on either probe, except for small changes in the directions of the two probes. So it seems reasonable to assume that there are no such aftereffects, and that both probes continue on their paths to the end of their cycles.

### Rule: Successive probe directions are roughly opposite

This is used to explain particle spin.

### Rule: Probe-probe connections are the sole cause of gravity

There are only three kinds of connections: probe-core, probe-probe, and core-core. Of these, probe-core connections obviously involve electromagnetic effects, and have nothing to do with gravity. Core-core effects are extremely local in nature, being limited to the charge radii of the two particles.

Probe-probe connections are all that is left, and there is nothing they could do but account for gravitational effects. That being the case, it is quite reasonable to assume that they all involve attractive aftereffects.

## APPENDIX

### TABLE 2-2
### Redshift versus Stellar Distances

Data from Tables 2, 3, and 4, Sandage, Astrophysical Journal, V. 178, pp. 6-10, arranged in order of increasing distance. The second column of the following table gives the corresponding table and line number from Sandage's paper. The value of $\alpha$ given in the next-to-last column is defined as $\alpha \equiv D/(1-F)$. The values $v_E$ in the last column are the adjusted velocities relative to Earth, in km/sec, using equation (2-7); positive values are away from Earth, negative one are toward Earth.

| $n$ | T-L | Cluster | $z$ | $V_C$ | $F$ | $D$ | $\alpha$ | $v_E$ |
|-----|-----|---------|-----|-------|-----|-----|----------|-------|
| 1  | 2-1  | VIRGO    | 0.00379 | 8.45  | 0.9962 | 0.0638 | 16.904 | 170.3 |
| 2  | 2-2  | FORNAX   | 0.00509 | 8.90  | 0.9949 | 0.0785 | 15.505 | 337.5 |
| 3  | 3-6  | 1332-33  | 0.0114  | 10.87 | 0.9887 | 0.1945 | 17.258 | 436.6 |
| 4  | 3-5  | 1245-41  | 0.0113  | 10.90 | 0.9888 | 0.1972 | 17.651 | 364.9 |
| 5  | 4-18 | 1060     | 0.0115  | 10.93 | 0.9886 | 0.2000 | 17.589 | 382.6 |
| 6  | 3-10 | 2318+07  | 0.0118  | 10.96 | 0.9883 | 0.2028 | 17.386 | 430.5 |
| 7  | 3-7  | 1400-33  | 0.0138  | 11.01 | 0.9864 | 0.2075 | 15.242 | 934.1 |
| 8  | 2-3  | PEG I    | 0.0128  | 11.20 | 0.9874 | 0.2265 | 17.918 | 339.7 |
| 9  | 2-6  | COMA     | 0.0222  | 11.51 | 0.9783 | 0.2612 | 12.027 | 2,570.7 |
| 10 | 2-5  | PERSEUS  | 0.0181  | 11.54 | 0.9822 | 0.2648 | 14.897 | 1,330.6 |
| 11 | 2-4  | 0122+3305 | 0.0170 | 11.70 | 0.9833 | 0.2851 | 17.055 | 684.4 |
| 12 | 4-4  | 262      | 0.0168  | 11.84 | 0.9835 | 0.3041 | 18.404 | 331.0 |
| 13 | 2-32 | 3C31     | 0.0169  | 11.90 | 0.9834 | 0.3126 | 18.809 | 230.3 |
| 14 | 4-11 | 569      | 0.0193  | 11.97 | 0.9811 | 0.3228 | 17.050 | 773.9 |
| 15 | 3-3  | 0153+05  | 0.0188  | 11.98 | 0.9815 | 0.3243 | 17.576 | 628.8 |
| 16 | 3-1  | 0036+03  | 0.0145  | 12.11 | 0.9857 | 0.3443 | 24.091 | -961.3 |
| 17 | 4-25 | 1367     | 0.0204  | 12.14 | 0.9800 | 0.3491 | 17.463 | 703.8 |
| 18 | 2-33 | 3C40     | 0.0180  | 12.24 | 0.9823 | 0.3656 | 20.675 | -253.6 |
| 19 | 4-24 | 1318     | 0.0189  | 12.33 | 0.9815 | 0.3810 | 20.542 | -247.8 |
| 20 | 4-33 | 2666     | 0.0273  | 12.37 | 0.9734 | 0.3881 | 14.605 | 2,119.1 |
| 21 | 3-9  | 2247+11  | 0.0268  | 12.40 | 0.9739 | 0.3935 | 15.077 | 1,883.0 |
| 22 | 2-36 | 3C338    | 0.0303  | 12.53 | 0.9706 | 0.4178 | 14.207 | 2,518.9 |
| 23 | 2-34 | 3C66     | 0.0215  | 12.69 | 0.9790 | 0.4498 | 21.369 | -551.5 |
| 24 | 4-30 | 2197     | 0.0322  | 12.71 | 0.9688 | 0.4539 | 14.551 | 2,514.5 |
| 25 | 4-29 | 2162     | 0.0318  | 13.00 | 0.9692 | 0.5188 | 16.832 | 1,387.0 |

**TABLE 2-2**
(continued)

| $n$ | T-L | Cluster | $z$ | $V_C$ | $F$ | $D$ | $\alpha$ | $v_E$ |
|---|---|---|---|---|---|---|---|---|
| 26 | 4-14 | 634 | 0.0266 | 13.03 | 0.9741 | 0.5260 | 20.300 | -234.1 |
| 27 | 2-35 | 3C465 | 0.0301 | 13.06 | 0.9708 | 0.5333 | 18.251 | 668.7 |
| 28 | 4-8 | 539 | 0.0267 | 13.11 | 0.9740 | 0.5457 | 20.985 | -511.3 |
| 29 | 2-7 | HERCULES | 0.0341 | 13.12 | 0.9670 | 0.5482 | 16.626 | 1,608.1 |
| 30 | 3-2 | 0131-36 | 0.0298 | 13.19 | 0.9711 | 0.5662 | 19.567 | 62.5 |
| 31 | 3-8 | 2152-69 | 0.0266 | 13.19 | 0.9741 | 0.5662 | 21.852 | -863.4 |
| 32 | 4-23 | 1314 | 0.0335 | 13.21 | 0.9676 | 0.5714 | 17.630 | 1,061.7 |
| 33 | 4-27 | 2147 | 0.0351 | 13.37 | 0.9661 | 0.6151 | 18.141 | 843.0 |
| 34 | 2-37 | 3C317 | 0.0351 | 13.38 | 0.9661 | 0.6180 | 18.224 | 797.6 |
| 35 | 4-9 | 548 | 0.0391 | 13.38 | 0.9624 | 0.6180 | 16.423 | 1,942.7 |
| 36 | 4-1 | A76 | 0.0377 | 13.38 | 0.9637 | 0.6180 | 17.010 | 1,540.4 |
| 37 | 2-8 | 2308+0720 | 0.0428 | 13.57 | 0.9590 | 0.6745 | 16.434 | 2,114.9 |
| 38 | 4-16 | 754 | 0.0537 | 13.58 | 0.9490 | 0.6776 | 13.296 | 5,171.0 |
| 39 | 3-4 | 0915-11 | 0.0522 | 13.58 | 0.9504 | 0.6776 | 13.659 | 4,736.4 |
| 40 | 4-15 | 671 | 0.0497 | 13.59 | 0.9527 | 0.6807 | 14.378 | 3,974.2 |
| 41 | 4-7 | 505 | 0.0543 | 13.61 | 0.9485 | 0.6870 | 13.340 | 5,180.9 |
| 42 | 4-21 | 1228 | 0.0344 | 13.65 | 0.9667 | 0.6998 | 21.043 | -674.1 |
| 43 | 4-26 | 1736 | 0.0431 | 13.73 | 0.9587 | 0.7261 | 17.572 | 1,400.0 |
| 44 | 4-19 | 1139 | 0.0376 | 13.75 | 0.9638 | 0.7328 | 20.222 | -293.1 |
| 45 | 4-12 | 576 | 0.0404 | 13.77 | 0.9612 | 0.7396 | 19.046 | 409.2 |
| 46 | 4-2 | 119 | 0.0387 | 13.79 | 0.9627 | 0.7464 | 20.034 | -165.4 |
| 47 | 4-20 | 1213 | 0.0287 | 13.80 | 0.9721 | 0.7499 | 26.877 | -3,150.1 |
| 48 | 4-6 | 376 | 0.0487 | 13.88 | 0.9536 | 0.7780 | 16.753 | 2,174.6 |
| 49 | 4-28 | 2152 | 0.0440 | 13.89 | 0.9579 | 0.7816 | 18.545 | 775.8 |
| 50 | 4-17 | 993 | 0.0530 | 13.93 | 0.9497 | 0.7961 | 15.817 | 3,108.4 |
| 51 | 2-11 | 0106-1536 | 0.0526 | 14.03 | 0.9500 | 0.8336 | 16.682 | 2,425.8 |
| 52 | 4-31 | 2319 | 0.0549 | 14.05 | 0.9480 | 0.8414 | 16.167 | 2,929.3 |
| 53 | 2-9 | 2322+1425 | 0.0440 | 14.08 | 0.9579 | 0.8531 | 20.241 | -356.8 |
| 54 | 4-22 | 1257 | 0.0339 | 14.08 | 0.9672 | 0.8531 | 26.017 | -3,270.9 |
| 55 | 4-3 | 147 | 0.0441 | 14.28 | 0.9578 | 0.9354 | 22.145 | -1,639.7 |

## TABLE 2-2
### (concluded)

| $n$ | T-L | Cluster | $z$ | $V_C$ | $F$ | $D$ | $\alpha$ | $v_E$ |
|---|---|---|---|---|---|---|---|---|
| 56 | 4-32 | 2657 | 0.0414 | 14.31 | 0.9602 | 0.9484 | 23.856 | -2,604.1 |
| 57 | 2-10 | 1145+5559 | 0.0516 | 14.39 | 0.9509 | 0.9840 | 20.053 | -244.0 |
| 58 | 4-13 | 592 | 0.0621 | 14.41 | 0.9415 | 0.9931 | 16.984 | 2,577.7 |
| 59 | 2-13 | 1239+1852 | 0.0718 | 14.52 | 0.9330 | 1.0447 | 15.594 | 4,447.6 |
| 60 | 4-10 | 553 | 0.0670 | 14.69 | 0.9372 | 1.1297 | 17.991 | 1,762.4 |
| 61 | 2-12 | 1024+1039 | 0.0649 | 14.81 | 0.9391 | 1.1939 | 19.590 | 123.8 |
| 62 | 2-15 | 0705+3506 | 0.0779 | 15.15 | 0.9277 | 1.3963 | 19.321 | 493.5 |
| 63 | 4-5 | 278 | 0.0904 | 15.19 | 0.9171 | 1.4223 | 17.155 | 3,493.1 |
| 64 | 2-14 | 1520+2754 | 0.0722 | 15.24 | 0.9327 | 1.4554 | 21.613 | -2,092.8 |
| 65 | 2-38 | M32-112 | 0.0825 | 15.33 | 0.9238 | 1.5170 | 19.905 | -223.4 |
| 66 | 2-19 | 1153+2341 | 0.1426 | 15.46 | 0.8752 | 1.6106 | 12.905 | 14,091.8 |
| 67 | 2-16 | 1513+0433 | 0.0944 | 15.50 | 0.9137 | 1.6405 | 19.019 | 1,030.8 |
| 68 | 2-18 | 1055+5702 | 0.1345 | 15.93 | 0.8814 | 1.9998 | 16.868 | 5,751.9 |
| 69 | 2-17 | 1431+3146 | 0.1312 | 16.06 | 0.8840 | 2.1231 | 18.306 | 2,819.3 |
| 70 | 2-27 | 0925+2044 | 0.1917 | 16.54 | 0.8391 | 2.6484 | 16.464 | 9,234.4 |
| 71 | 2-21 | 1534+3749 | 0.1532 | 16.55 | 0.8672 | 2.6606 | 20.027 | -710.5 |
| 72 | 2-22 | 0025+2223 | 0.1594 | 16.58 | 0.8625 | 2.6976 | 19.621 | 268.9 |
| 73 | 2-20 | 1641+1327 | 0.1499 | 16.61 | 0.8696 | 2.7351 | 20.981 | -2,863.1 |
| 74 | 2-39 | 3C219 | 0.1745 | 16.88 | 0.8514 | 3.0973 | 20.847 | -2,981.6 |
| 75 | 2-26 | 1304+3110 | 0.1831 | 16.90 | 0.8452 | 3.1259 | 20.198 | -1,294.5 |
| 76 | 2-23 | 1228+1050 | 0.1651 | 16.90 | 0.8583 | 3.1259 | 22.059 | -5,961.1 |
| 77 | 2-25 | 1309-0105 | 0.1745 | 16.93 | 0.8514 | 3.1694 | 21.332 | -4,299.8 |
| 78 | 2-29 | 0855+0321 | 0.2018 | 17.07 | 0.8321 | 3.3805 | 20.132 | -1,243.8 |
| 79 | 2-40 | 3C28 | 0.1959 | 17.10 | 0.8362 | 3.4275 | 20.924 | -3,599.3 |
| 80 | 2-24 | 0138+1832 | 0.1730 | 17.17 | 0.8525 | 3.5398 | 24.001 | -11,658.7 |
| 81 | 2-28 | 1253+4422 | 0.1979 | 17.66 | 0.8348 | 4.4359 | 26.851 | -23,061.6 |
| 82 | 2-31 | 0024+1654 | 0.38 | 18.34 | 0.7246 | 6.0671 | 22.033 | -13,899.4 |
| 83 | 2-41 | 3C295 | 0.461 | 18.63 | 0.6845 | 6.9339 | 21.975 | -16,612.7 |
| 84 | 2-30 | 1447+2617 | 0.36 | 18.77 | 0.7353 | 7.3957 | 27.939 | -52,820.2 |

# CHAPTER 3:  THE SPEED OF LIGHT

In 1905, in the development of special relativity, Einstein decided that light was propagated by wave packets (later renamed photons).  He could see no reason that a wave packet would slow down as it traveled through space.  Hence he assumed that the speed of light is constant.

However, as shown in Chapter 1, the concept of a photon is unreasonable.  Hence Einstein's justification for assuming that the speed of light is constant is invalid.

On the other hand, new research shows that his assumption is mostly valid, but for a very different reason.  Recall that only electrons, protons, and their antiparticles have a core-probe structure.  Hence for this discussion, the word *particle* refers either to an electron, a proton, or one of their antiparticles.

Two numbers are associated with the probe of a particle: $M$, the maximum distance that the probe tip can travel from its core, and $s$, the time to complete one cycle.  (Since these cycles are not periodic in the sense that the paths are repeated periodically, a cycle is defined as the time for one probe to leave the core and then return to it.)  The latter number is the inverse of particle frequency $\nu$ : $s = 1/\nu$ .  Thus for an electron at rest, $\nu_e = 1.2355 \times 10^{20}$ cycles per second, hence $s_e = 8.0939 \times 10^{-21}$ seconds.  For a proton at rest, $\nu_p = 2.2687 \times 10^{23}$ cycles per second, hence $s_p = 4.4078 \times 10^{-24}$ seconds.

Using data from Table 2-2, the value of $M_e$ has been estimated at $1.8673 \times 10^{26}$ meters, which is about 19.74 billion light years.  Since the frequency of a proton is about 1,836 times that of an electron, it follows that the estimate of $M$ for a proton is about $1/1836$ that of an electron, thus $M_p = 1.0170 \times 10^{23}$, or about 10.75 million light years.

Thus the tip of an electron probe, if it does not encounter any probes or cores on its journey, travels out about $1.8673 \times 10^{26}$ meters and back in $8.0939 \times 10^{-21}$ seconds.  This is about $4.6 \times 10^{46}$ meters per second.  Well, the universe seems to have a rule that any object which wants to travel at a local speed greater than 299,792,458 meters per second must go on an alternate path through the universe at an apparent speed of 299,792,458 meters per second, and into the future at a rate of clock time equal to the distance traveled so far divided by 299,792,458.

A similar calculation can be made for proton probes.

Thus it is evident that probes through space are always traveling at 299,792,458 meters per second, except when they are nearly to their maximum distance, at which point their local speed drops below that value as they finally come to a stop.  Thus Einstein's assumption is essentially correct.

# CHAPTER 4:  TIRED LIGHT

Cosmology has evolved over the centuries.  Two thousand years ago, the prevailing cosmology was geocentric.  It was then obvious to most people that the Moon, the Sun, and the stars revolved around Earth.

In the 16th century, Nicolaus Copernicus made the case for a heliocentric cosmology: Everything in the solar system revolved around the Sun.

Today, a further evolution in cosmology has been forced upon us.  Astronomical observations have indicated that stars in galaxies other than our own are moving so fast that gravity cannot hold them together.  And the farther they are from Earth, the faster those stars are moving.

To explain the reason for this, scientists have come up with the idea that there must be dark matter (or dark energy) holding them together.

Thus our own galaxy (the Milky Way) seems to be in a favored position in the universe.  It is the only galaxy, or at least one of a tiny number out of the estimated two trillion galaxies in the universe, which does not seem to require any dark matter.  Thus cosmology now seems bent on the idea that the universe somehow favors our galaxy, if not being centered on it.

It seems rash to think that our galaxy is so special.  But what about the huge number of observations that have led to this situation?  How can so many astronomers be wrong?

A simple answer to that question was suggested almost a century ago by the astronomer Fritz Zwicky.  His suggestion was that photons encountered things that slowed them down as they passed through space.  This idea was later termed *tired light*.

But nothing was ever found which could accomplish that.

However, with the concept of probes, another idea is suggested, namely, that at the beginning of a cycle, the mass of the probe tip equals the measured mass of the particle.  As the probe tip extends outward, it leaves a faint trail of mass behind it.  Think of it as like a yo-yo flying out.  The farther it goes, the more mass is left behind it.  Consequently, when the probe tip makes a connection with another probe or another core, it has only the remaining mass to use for the contact.

Thus an electron probe tip, after going a distance $D$ toward its maximum distance $M_e$, has mass $m$ equal to its starting mass $m_e$ times $1-\dfrac{D}{M_e}$ left to use in a contact.  The value of $M_e$ was estimated in Chapter 2 to be about $1.8673 \times 10^{26}$ meters (19.74 billion light years).  That is,

(4-1) $\qquad m = m_e \left( 1 - \dfrac{D}{M_e} \right).$

Note: In Chapter 2 it was found that the velocity $v$ of the probe tip at that distance is $v = c \sqrt{1 - \dfrac{D}{M_e}}$ .

Hence on its outward journey, the mass and velocity of the probe tip satisfy the relation $\dfrac{v^2}{c^2} = \dfrac{m}{m_e}$ .

It is more useful to work in terms of wavelength $\lambda$ than mass. By the usual formulas,

$$mc^2 = E = ch/\lambda,$$

and so

$$m = h/c\lambda,$$
$$\lambda = h/mc.$$

Hence equation (4-1) becomes

(4-2) $\qquad \lambda = \lambda_e \left( 1 - \dfrac{D}{M_e} \right)^{-1}.$

Let $\lambda_e$ be the wavelength of a probe at the beginning of a cycle on a star which is a distance $D$ from Earth, and let $\lambda_{obs}$ be the wavelength of that probe when it reaches Earth. Then by equation (4-2),

$$\lambda_{obs} = \lambda_e \left( 1 - \dfrac{D}{M_e} \right)^{-1}.$$

Then $\lambda_e$ can be estimated by rearranging the above equation to get an estimate $\lambda_{est}$ of $\lambda_e$:

(4-3) $\qquad \lambda_{est} = \lambda_{obs} \left( 1 - \dfrac{D}{M_e} \right).$

(This is an estimate because $M_e$ is an estimate.)

It is expected that when this formula is used before calculating the redshift of remote stars or galaxies, the results will show that the corresponding velocities are low enough that gravity can hold remote galaxies together. Hence the need for dark matter or dark energy vanishes.

After applying equation (4-3), then calculate the redshift $z$ from

$$z = \frac{\lambda_{\text{est}} - \lambda_{\text{std}}}{\lambda_{\text{std}}}.$$

where $\lambda_{\text{std}}$ is the expected value for $\lambda$ when the remote object is at a constant distance from Earth (neither receding nor approaching).

These steps lead to a value of $z$ which compensates for the loss of energy of the probe on its outward journey.

Table 4-1 at the end of this chapter uses data from Tables 2, 3, and 4, of the paper by Allan Sandage in the Astrophysical Journal, V. 178, pp. 6-29. The data are arranged in order of increasing distance from Earth.

In that paper, Sandage refers to several manuscripts in which various colleagues made calculations in support of the paper. I presume that these manuscripts included calculations of the redshift values used in the paper. Such calculations would have taken the form

$$z = \frac{\lambda_{\text{obs}} - \lambda_{\text{std}}}{\lambda_{\text{std}}}$$

where $z$ is the redshift, $\lambda_{\text{obs}}$ is the observed wavelength, and $\lambda_{\text{std}}$ is the standard wavelength for that spectral line.

Equation (4-3) should be applied to $\lambda_{\text{obs}}$ before the above redshift formula is used. But the values of $\lambda_{\text{obs}}$, presumably in those manuscripts, were not available to me at the time of writing this book.

In lieu of those values, I took three arbitrary wavelengths, namely 3819Å, 4144Å, and 4481Å. To estimate $\lambda_{\text{obs}}$, I solved the equation

$$\lambda_{\text{obs}} = \lambda_{\text{std}} (z+1)$$

for each of those three standard wavelengths, and for each data point in Table 4-1.

For each of the resulting estimated values of $\lambda_{obs}$, I applied equation (4-3) to get a corrected value:

$$\lambda_{est} = \lambda_{obs}\left(1 - D/M_e\right),$$

where $D$ is the distance of the cluster from Earth as given in Table 4-1, and $M_e$ is the estimated maximum distance that light can travel.

Each such corrected wavelength $\lambda_{est}$ was then used to calculate the corrected redshift for that cluster:

$$z_{corr} = \frac{\lambda_{est} - \lambda_{std}}{\lambda_{std}}.$$

It turned out that the results were the same for each of the three wavelengths used for $\lambda_{std}$. Therefore, only the results for the middle wavelength, 4144Å, are reported in Table 4-1.

A point by point plot of the raw redshift values (those from Sandage's paper) is shown below, followed by a plot of the corrected redshift values.

**Figure 4-1**
**Raw $z$**

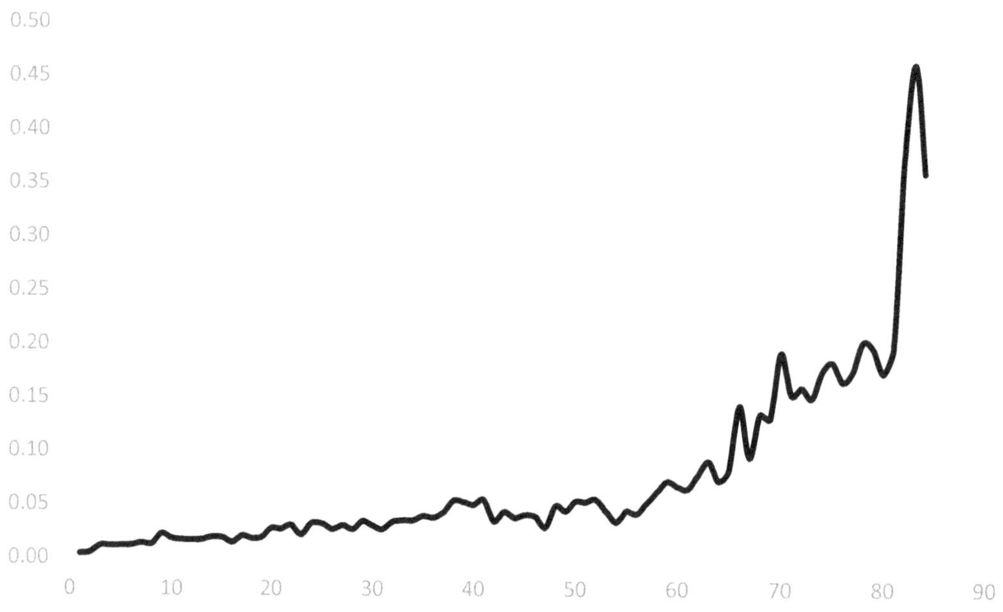

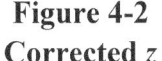

**Figure 4-2**
**Corrected $z$**

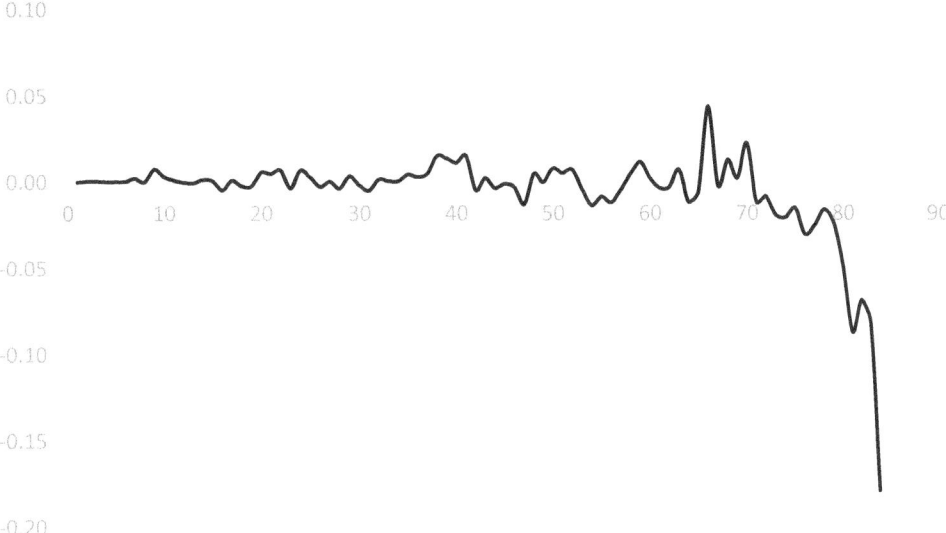

Whereas the raw $z$ values increase almost exponentially, the corrected values wobble about 0, at least for the first 70 points. After that, the data are not sufficiently accurate to show any reasonable trend. Of course, the distances from Earth of points 71 to 84 are greater than 2.8 billion light years.

Thus equation (4-3) reduces most of Sandage's data to roughly zero values. Since that formula was derived from theoretic considerations, and without any data, I assert that the data fit proves that the formula is reasonable, and that the behavior of an electron as described in earlier chapters is also justified. The exceptional values are associated with extremely large distances, which may not be accurate.

In view of the above, it seems reasonable to say that stars in remote galaxies are not moving away from Earth at high velocities, but instead are moving at modest velocities either toward or away from Earth. Thus, the universe is not expanding, and there is no reason to think that dark matter or dark energy exist.

Table 4-1 follows. It is a modification of Table 2-2 from the appendix to Chapter 2. The first line contains the values of the constant $M_e$ (in units of $10^{25}$ meters), and the arbitrary wavelength used for $\lambda_{std}$ (in Ångströms). From there on, the columns of the table consist of (1) the point numbers, (2) the redshift values from Sandage's paper, (3) the distances of the points from Earth (also in units of $10^{25}$ meters), derived from Sandage's paper, (4) the observed wavelength as inferred from a formula above, (5) the estimated wavelength of $\lambda_{dop}$, applying equation (4-3), and (6) the corrected redshift values.

## TABLE 4-1
### Redshift versus Distance

| $n$ | $z$ | $D$ | $M_e$ 18.673 | | $\lambda_{std}$ 4144 | |
|---|---|---|---|---|---|---|
| | | | $\lambda_{obs}$ | $\lambda_{est}$ | $z_{corr}$ | |
| 1 | 0.00379 | 0.0638 | 4160 | 4145 | 0.00036 |
| 2 | 0.00509 | 0.0785 | 4165 | 4148 | 0.00086 |
| 3 | 0.0114 | 0.1945 | 4191 | 4148 | 0.00087 |
| 4 | 0.0113 | 0.1972 | 4191 | 4147 | 0.00062 |
| 5 | 0.0115 | 0.2000 | 4192 | 4147 | 0.00067 |
| 6 | 0.0118 | 0.2028 | 4193 | 4147 | 0.00081 |
| 7 | 0.0138 | 0.2075 | 4201 | 4155 | 0.00253 |
| 8 | 0.0128 | 0.2265 | 4197 | 4146 | 0.00051 |
| 9 | 0.0222 | 0.2612 | 4236 | 4177 | 0.00790 |
| 10 | 0.0181 | 0.2648 | 4219 | 4159 | 0.00366 |
| 11 | 0.0170 | 0.2851 | 4214 | 4150 | 0.00147 |
| 12 | 0.0168 | 0.3041 | 4214 | 4145 | 0.00024 |
| 13 | 0.0169 | 0.3126 | 4214 | 4143 | -0.00012 |
| 14 | 0.0193 | 0.3228 | 4224 | 4151 | 0.00168 |
| 15 | 0.0188 | 0.3243 | 4222 | 4149 | 0.00111 |
| 16 | 0.0145 | 0.3443 | 4204 | 4127 | -0.00421 |
| 17 | 0.0204 | 0.3491 | 4229 | 4149 | 0.00132 |
| 18 | 0.0180 | 0.3656 | 4219 | 4136 | -0.00193 |
| 19 | 0.0189 | 0.3810 | 4222 | 4136 | -0.00189 |
| 20 | 0.0273 | 0.3881 | 4257 | 4169 | 0.00595 |
| 21 | 0.0268 | 0.3935 | 4255 | 4165 | 0.00516 |
| 22 | 0.0303 | 0.4178 | 4270 | 4174 | 0.00725 |
| 23 | 0.0215 | 0.4498 | 4233 | 4131 | -0.00311 |
| 24 | 0.0322 | 0.4539 | 4277 | 4173 | 0.00711 |
| 25 | 0.0318 | 0.5188 | 4276 | 4157 | 0.00313 |
| 26 | 0.0266 | 0.5260 | 4254 | 4134 | -0.00232 |
| 27 | 0.0301 | 0.5333 | 4269 | 4147 | 0.00068 |
| 28 | 0.0267 | 0.5457 | 4255 | 4130 | -0.00330 |
| 29 | 0.0341 | 0.5482 | 4285 | 4160 | 0.00374 |

**TABLE 4-1**
**Redshift versus Distance**
(continued)

| $n$ | $z$ | $D$ | $\lambda_{obs}$ | $\lambda_{est}$ | $z_{corr}$ |
|---|---|---|---|---|---|
| 30 | 0.0298 | 0.5662 | 4267 | 4138 | -0.00143 |
| 31 | 0.0266 | 0.5662 | 4254 | 4125 | -0.00453 |
| 32 | 0.0335 | 0.5714 | 4283 | 4152 | 0.00187 |
| 33 | 0.0351 | 0.6151 | 4289 | 4148 | 0.00100 |
| 34 | 0.0351 | 0.6180 | 4289 | 4147 | 0.00084 |
| 35 | 0.0391 | 0.6180 | 4306 | 4164 | 0.00471 |
| 36 | 0.0377 | 0.6180 | 4300 | 4158 | 0.00336 |
| 37 | 0.0428 | 0.6745 | 4321 | 4165 | 0.00513 |
| 38 | 0.0537 | 0.6776 | 4367 | 4208 | 0.01546 |
| 39 | 0.0522 | 0.6776 | 4360 | 4202 | 0.01402 |
| 40 | 0.0497 | 0.6807 | 4350 | 4191 | 0.01143 |
| 41 | 0.0543 | 0.6870 | 4369 | 4208 | 0.01551 |
| 42 | 0.0344 | 0.6998 | 4287 | 4126 | -0.00437 |
| 43 | 0.0431 | 0.7261 | 4323 | 4155 | 0.00254 |
| 44 | 0.0376 | 0.7328 | 4300 | 4131 | -0.00312 |
| 45 | 0.0404 | 0.7396 | 4311 | 4141 | -0.00081 |
| 46 | 0.0387 | 0.7464 | 4304 | 4132 | -0.00282 |
| 47 | 0.0287 | 0.7499 | 4263 | 4092 | -0.01261 |
| 48 | 0.0487 | 0.7780 | 4346 | 4165 | 0.00501 |
| 49 | 0.0440 | 0.7816 | 4326 | 4145 | 0.00030 |
| 50 | 0.0530 | 0.7961 | 4364 | 4178 | 0.00811 |
| 51 | 0.0526 | 0.8336 | 4362 | 4167 | 0.00561 |
| 52 | 0.0549 | 0.8414 | 4372 | 4175 | 0.00737 |
| 53 | 0.0440 | 0.8531 | 4326 | 4129 | -0.00370 |
| 54 | 0.0339 | 0.8531 | 4284 | 4089 | -0.01334 |
| 55 | 0.0441 | 0.9354 | 4327 | 4110 | -0.00820 |
| 56 | 0.0414 | 0.9484 | 4316 | 4096 | -0.01149 |
| 57 | 0.0516 | 0.9840 | 4358 | 4128 | -0.00382 |
| 58 | 0.0621 | 0.9931 | 4401 | 4167 | 0.00561 |
| 59 | 0.0718 | 1.0447 | 4442 | 4193 | 0.01184 |

## TABLE 4-1
### Redshift versus Distance
(concluded)

| $n$ | $z$ | $D$ | $\lambda_{obs}$ | $\lambda_{est}$ | $z_{corr}$ |
|---|---|---|---|---|---|
| 60 | 0.0670 | 1.1297 | 4422 | 4154 | 0.00245 |
| 61 | 0.0649 | 1.1939 | 4413 | 4131 | -0.00319 |
| 62 | 0.0779 | 1.3963 | 4467 | 4133 | -0.00270 |
| 63 | 0.0904 | 1.4223 | 4519 | 4174 | 0.00735 |
| 64 | 0.0722 | 1.4554 | 4443 | 4097 | -0.01137 |
| 65 | 0.0825 | 1.5170 | 4486 | 4121 | -0.00544 |
| 66 | 0.1426 | 1.6106 | 4735 | 4327 | 0.04405 |
| 67 | 0.0944 | 1.6405 | 4535 | 4137 | -0.00175 |
| 68 | 0.1345 | 1.9998 | 4701 | 4198 | 0.01300 |
| 69 | 0.1312 | 2.1231 | 4688 | 4155 | 0.00258 |
| 70 | 0.1917 | 2.6484 | 4938 | 4238 | 0.02268 |
| 71 | 0.1532 | 2.6606 | 4779 | 4098 | -0.01111 |
| 72 | 0.1594 | 2.6976 | 4805 | 4110 | -0.00809 |
| 73 | 0.1499 | 2.7351 | 4765 | 4067 | -0.01853 |
| 74 | 0.1745 | 3.0973 | 4867 | 4060 | -0.02031 |
| 75 | 0.1831 | 3.1259 | 4903 | 4082 | -0.01495 |
| 76 | 0.1651 | 3.1259 | 4828 | 4020 | -0.02994 |
| 77 | 0.1745 | 3.1694 | 4867 | 4041 | -0.02485 |
| 78 | 0.2018 | 3.3805 | 4980 | 4079 | -0.01577 |
| 79 | 0.1959 | 3.4275 | 4956 | 4046 | -0.02361 |
| 80 | 0.1703 | 3.5398 | 4861 | 3939 | -0.04936 |
| 81 | 0.1979 | 4.4359 | 4964 | 3785 | -0.08667 |
| 82 | 0.3800 | 6.0671 | 5719 | 3861 | -0.06838 |
| 83 | 0.4610 | 6.9339 | 6054 | 3806 | -0.08152 |
| 84 | 0.3600 | 7.3957 | 5636 | 3404 | -0.17865 |

## CHAPTER 5: PARTICLE INTERACTIONS INVOLVING DIRECT CONTACT

### Neutrons

The neutron is a close combination of an electron and a proton. A close combination of two particles (or equivalently, two particles in direct contact) means that the distance between the cores of the particles remains less than the charge radius of the proton. According to the current Wikipedia article on Proton, this is a variable quantity with a root mean square value between $0.84 \times 10^{-15}$ and $0.87 \times 10^{-15}$ meters.

There is quantity called the classical electron radius, which may or may not be the equivalent of the proton charge radius. Its value is $2.82 \times 10^{-15}$ meters. Hence if the distance between the two cores is less than the proton charge radius, it is certainly within the classical electron radius.

The result of being this close is that every time the probe of either particle returns to its core, it will form a connection with the core of the other particle. Each such connection yields an attractive impulse between the two cores. This connection is an example of the strong nuclear force, which generally keeps the two cores within the proton charge radius. However, there are factors at work in this process which can lead to a situation where the cores separate by more than that charge radius, leading to the decay of the neutron. These factors are discussed next.

One of these factors was discussed in Chapter 4: The masses of the electron and the proton probe tips vary with time. During a cycle, the probe tip of the electron moves from its core out to a distance of $M_e$, where $M_e$ is estimated to be about $1.8673 \times 10^{26}$ meters. The corresponding value $M_p$ for a proton probe tip is $1.0170 \times 10^{23}$ meters.

Thus when the probe tip of either particle returns to its core, the connection formed between two cores has the full mass of one particle connected to the mass of the core of the other particle. At present, the value of either core mass is unknown.

The other factor is that the timing of the connections is irregular. There will be over 700 connections due to the proton probe between every two connections due to the electron probe. The precise numbers depend on the current velocities of the cores, which change as the result of each connection.

The appendix to this chapter contains a guide to completing the model of the neutron.

### Elementary Particles

The standard model of physics seems to have two main ideas regarding other elementary particles such as muons, pions, kaons, etc. First, such a particle consists of a bundle of energy (not of individual particles). Second, the bundle is held together by the strong nuclear force.

Connection theory has a different interpretation of elementary particles. First, an elementary particle consists of a close combination of primitive particles. Second, these primitive particles are held together by the strong nuclear force, constrained by the conservation laws of mass/energy and momentum. The strong nuclear force in such a case is generated by the connections formed when the probe of any constituent particle returns to its core. At such a time, that probe will form a connection with the core of every other constituent particle. The accumulation of these connections yields a form of the strong nuclear force particular to that particle. Thus the strong nuclear force holding an elementary particle together is different for different elementary particles.

### Decay of Elementary Particles

It has been observed that an elementary particle decays into smaller elementary particles, which then decay into yet smaller particles, ending with primitive particles. At each step in the decay, energy is released in the form of photons, neutrinos, and/or antineutrinos.

There are two different causes of elementary particle decay: *internal decay* and *escape*. Internal decay can occur if the elementary particle contains both a particle and its antiparticle, such as an electron and a positron. Internal decay occurs when these two particles annihilate each other.

The constituent particles within an elementary particle $A$ are battered about within $A$ every time the probe of a constituent particle returns to its core. At such an evert, that probe forms a connection with the core of every other constituent particle. Between every such cycle and the next, these constituents move in straight lines. Escape is where one of these particles, say $P$, between one cycle and the next, moves outside the charge radii of the other constituents, so that when the next such event occurs, the particle $P$ does not experience the resulting connection, and thus escapes from $A$.

One objection to the connection-theoretic model is that it assumes that a primitive particle and its antiparticle can exist together in the elementary particle. The response to this objection is that in the most extreme case (para-positronium), which consists of an electron and a positron, it takes typically about 125 picoseconds for these two particles to annihilate each other. And many elementary particles have a half-life of a lot more than that. Hence it is reasonable to suppose that in an elementary particle with a half-life measured in microseconds, an electron and a positron could coexist quite comfortably for a short while.

## APPENDIX
### A Guide to Completing the Model of the Neutron

It was stated previously that the effective mass of the core of a particle when its probe is away from the core is unknown. However, it seems reasonable to assume that the core itself has a mass independent of the mass of the probe, and that when the probe is at the core, then the effective mass is equal to the core mass plus the probe mass.

Hence define the following symbols:

$m_{ec}$     the mass of the core of the electron,

$m_{ep}$     the mass of the electron probe,

$m_{pc}$     the mass of the core of the proton,

$m_{pp}$     the mass of the proton probe.

Then $m_{ep} + m_{ec}$ equals the mass of the electron when its probe is at its core. Similarly, $m_{pp} + m_{pc}$ equals the mass of the proton when its probe is at its core.

It might be thought that $m_{ep} + m_{ec}$ equals the measured mass of the electron. But evidence is that the measured mass of the electron is actually equal to $m_{ep}$. The reason is that the effect of $m_{ec}$ only plays a role for tiny fraction of a zeptosecond—too small to be noticed. A similar comment applies to the proton. Thus the two masses $m_{ec}$ and $m_{pc}$ remain to be determined.

In case E, when the electron probe returns to its core, the resulting connection with the proton core involves the two masses $m_{ep} + m_{ec}$ and $m_{pc}$, whereas in case P, when the proton probe returns to its core, the resulting connection involves the two masses $m_{pp} + m_{pc}$ and $m_{ec}$. (The case that both probes return at the same instant is so unlikely that it will be ignored.)

Suppose at the time of case E or case P that the position and velocity values of the particles before and after the connection are as shown in the following table:

| Particle | Case E | | Case P | |
|---|---|---|---|---|
| | Position | Velocity | Position | Velocity |
| Electron before | $x_{eb}^{E}$ | $v_{eb}^{E}$ | $x_{eb}^{P}$ | $v_{eb}^{P}$ |
| Electron after | $x_{ea}^{E}$ | $v_{ea}^{E}$ | $x_{ea}^{P}$ | $v_{ea}^{P}$ |
| Proton before | $x_{pb}^{E}$ | $v_{pb}^{E}$ | $x_{pb}^{P}$ | $v_{pb}^{P}$ |
| Proton after | $x_{pa}^{E}$ | $v_{pa}^{E}$ | $x_{pa}^{P}$ | $v_{pa}^{P}$ |

The basic equations governing the changes from the before positions and velocities to the after ones are due to the conservation of mass/energy:

(5-1)  $\qquad \gamma_{11}m_{11} + \gamma_{21}m_{21} = \gamma_{12}m_{12} + \gamma_{22}m_{22}$ ,

and the conservation of momentum:

(5-2)  $\qquad \gamma_{11}m_{11}v_{11}\vec{u}_{11} + \gamma_{21}m_{21}v_{21}\vec{u}_{21} = \gamma_{1}m_{12}v_{12}\vec{u}_{12} + \gamma_{22}m_{22}v_{22}\vec{u}_{22}$ .

In these equations, the first subscript identifies the particle, and the value of the second is 1 for before the event and 2 for after it. Also,

$\qquad v_{ij}$ is the velocity of the concerned particle,

$\qquad m_{ij}$ is the current mass of that particle,

$\qquad \gamma_{ij}$ is the symbol for the relativity factor of the that particle: $\gamma_{ij} = 1 \big/ \sqrt{1 - \left( v_{ij}/c \right)^{2}}$ ,

$\qquad \vec{u}_{ij}$ is the position vector of that particle.

However, for this situation, it is assumed that the particles move back and forth along a single axis. Therefore, in place of the vector $\vec{u}_{ij}$ , it suffices to use the scalar $x_{ij}^{k}$ for that purpose. Furthermore, the particle positions do not change during the connection, nor do the particle masses, apart from the relativity factors. Consequently, equations (5-1) and (5-2) are replaced by

(5-3)  $\qquad \gamma_{11}m_{11} + \gamma_{21}m_{21} = \gamma_{12}m_{11} + \gamma_{22}m_{21}$

and

(5-4)  $\qquad \gamma_{11}m_{11}v_{11}x_{11}^{k} + \gamma_{21}m_{21}v_{21}x_{21}^{l} = \gamma_{12}m_{11}v_{12}x_{11}^{k} + \gamma_{22}m_{21}v_{22}x_{21}^{l}$ ,

38

where $(k,l) = (E,P)$ or $(P,E)$.

Thus the equations for Case E are

(5-5)   $$\gamma_{eb}^{E}\left(m_{ep}+m_{ec}\right)+\gamma_{pb}^{E}m_{pc}=\gamma_{ea}^{E}\left(m_{ep}+m_{ec}\right)+\gamma_{pa}^{E}m_{pc}$$

and

(5-6)   $$\gamma_{eb}^{E}\left(m_{ep}+m_{ec}\right)v_{eb}^{E}x_{eb}^{E}+\gamma_{pb}^{E}m_{pc}v_{pb}^{E}x_{pb}^{E}=\gamma_{ea}^{E}\left(m_{ep}+m_{ec}\right)v_{ea}^{E}x_{eb}^{E}+\gamma_{pa}^{E}m_{pc}v_{pa}^{E}x_{pb}^{E}.$$

The equations for Case P are

(5-7)   $$\gamma_{pb}^{P}\left(m_{pp}+m_{pc}\right)+\gamma_{eb}^{P}m_{ec}=\gamma_{pa}^{P}\left(m_{pp}+m_{pc}\right)+\gamma_{ea}^{P}m_{ec}$$

and

(5-8)   $$\gamma_{pb}^{P}\left(m_{pp}+m_{pc}\right)v_{pb}^{P}x_{pb}^{P}+\gamma_{eb}^{P}m_{ec}v_{eb}^{P}x_{eb}^{P}=\gamma_{pa}^{P}\left(m_{pp}+m_{pc}\right)v_{pa}^{P}x_{pb}^{P}+\gamma_{ea}^{P}m_{ec}v_{ea}^{P}x_{eb}^{P}.$$

### Solving These Equations

Although the above equations may appear formidable and complex, they can be solved by iteration. The derivation of the procedure takes several pages, but the steps are reasonably easy to follow, and the resulting procedure is fairly simple

The derivation starts with equation (5-4), to reduce the complexity of what follows. From that equation, it can be seen that any connection between the proton and the electron involves the exchange of an impulse between them. Thus, the impulse which acts on the first particle (the difference between its linear momentum before and after the interaction) is

$$I_{1}=\gamma_{12}m_{11}v_{12}x_{11}^{k}-\gamma_{11}m_{11}v_{11}x_{11}^{k},$$

and the impulse which acts on the second particle is

$$I_{2}=\gamma_{22}m_{21}v_{22}x_{21}^{l}-\gamma_{21}m_{21}v_{21}x_{21}^{l}.$$

From equation (5-4), it follows that $I_{1}+I_{2}=0$, so $I_{2}=-I_{1}$. Thus the two particles exchange an impulse with magnitude $I=\left|I_{1}\right|=\left|I_{2}\right|$.

A combination of equations (5-3) and (5-4), with a slight rearrangement, yields the basic equations that define the results of a connection between these two particles:

$$\gamma_{12}m_{11} + \gamma_{22}m_{21} = \gamma_{11}m_{11} + \gamma_{21}m_{21}$$

(5-9)
$$\gamma_{12}m_{11}v_{12}x_{11}^k = \gamma_{11}m_{11}v_{11}x_{11}^k + I_1$$

$$\gamma_{22}m_{21}v_{22}x_{21}^l = \gamma_{21}m_{21}v_{21}x_{21}^l - I_1.$$

To solve these equations for the unknowns $I_1$, $v_{12}$, and $v_{22}$, thus yielding $\gamma_{12}$ and $\gamma_{22}$, is not difficult. Start with the second and third equations of (5-9) and square each one (that is, take the scalar product of each side by itself). Then divide each square by the square of the mass and the square of the position. The results are

(5-10)
$$\gamma_{12}^2 v_{12}^2 = \gamma_{11}^2 v_{11}^2 + \frac{2\gamma_{11}v_{11}I_1}{m_{11}x_{11}^k} + \frac{I_1^2}{m_{11}^2 \left(x_{11}^k\right)^2}$$

$$\gamma_{22}^2 v_{22}^2 = \gamma_{21}^2 v_{21}^2 - \frac{2\gamma_{21}v_{21}I_1}{m_{21}x_{21}^l} + \frac{I_1^2}{m_{21}^2 \left(x_{21}^l\right)^2}.$$

Note that except for $I_1$, the right sides of these equations consist of known quantities.

Next, define two intermediate variables $p$ and $q$ equal to the right sides of the equations of (5-10) each divided by $c^2$:

(5-11)
$$p = \frac{\gamma_{11}^2 v_{11}^2}{c^2} + \frac{2\gamma_{11}v_{11}I_1}{m_{11}c^2 x_{11}^k} + \frac{I_1^2}{m_{11}^2 c^2 \left(x_{11}^k\right)^2}$$

$$q = \frac{\gamma_{21}^2 v_{21}^2}{c^2} - \frac{2\gamma_{21}v_{21}I_1}{m_{21}c^2 x_{21}^l} + \frac{I_1^2}{m_{21}^2 c^2 \left(x_{21}^l\right)^2}.$$

(Thus both $p$ and $q$ are positive.) Then by equation (5-10),

$$p = \frac{\gamma_{12}^2 v_{12}^2}{c^2} = \frac{v_{12}^2/c^2}{1 - v_{12}^2/c^2} = \frac{1}{1 - v_{12}^2/c^2} - 1 = \gamma_{12}^2 - 1$$

$$q = \frac{\gamma_{22}^2 v_{22}^2}{c^2} = \frac{v_{22}^2/c^2}{1 - v_{22}^2/c^2} = \frac{1}{1 - v_{22}^2/c^2} - 1 = \gamma_{22}^2 - 1,$$

and so

$$\gamma_{12} = \sqrt{1 + p}$$

$$\gamma_{22} = \sqrt{1 + q}$$

and

$$v_{12} = c\sqrt{\frac{p}{1+p}}$$

(5-12)

$$v_{22} = c\sqrt{\frac{q}{1+q}}.$$

Thus

$$\gamma_{12}v_{12} = c\sqrt{p}$$

$$\gamma_{22}v_{22} = c\sqrt{q}.$$

The next step is to see if the quantities thus obtained satisfy equations (5-3) and (5-4). In other words, does there exist a value of $I_1$ for which

(5-13)
$$\gamma_{11}m_{11} + \gamma_{21}m_{21} = m_{11}\sqrt{1+p} + m_{21}\sqrt{1+q}$$
$$\gamma_{11}m_{11}v_{11}x_{11}^k + \gamma_{21}m_{21}v_{21}x_{21}^l = cm_{11}x_{11}^k\sqrt{p} + cm_{21}x_{21}^l\sqrt{q}$$ ?

It can easily be seen that there is always the solution $I_1 = 0$, which means that nothing changes. In a lot of cases, there is also a nonzero solution $I_1$, which leads to changes in the energy and velocity of the two particles. At a guess, nature chooses the nonzero $I_1$ if one exists.

A possible approach to finding a solution consists of three steps. The first step is to assume values for $m_{ec}$ and $m_{pc}$.

The second step is to use these values with a wide range of possible values of $I_1$ in the two forms of equation (5-13). Examine the results to see if there is any value of $I_1$ for which both forms of equation (5-13) are satisfied. For each such $I_1$, proceed to the third step.

The third step is to use the value found for $I_1$ to run a simulation of the behavior of the neutron over a few thousand cycles. Use the resulting values to calculate the average velocities of the particles. If these average velocities yield the measured mass of the neutron, then the assumed values of $m_{ec}$ and $m_{pc}$ are reasonable.

Though it is easy to describe the procedure, it seems likely that it would take a substantial amount of work to accomplish it. At age 87, the present author lacks the energy to tackle such a task. He hopes that some other researcher will take it on!

## CHAPTER 6: GRAVITY AND PERIHELIA PRECESSIONS

### Gravity

Consider two cores that are emitting probes. Some of these probes will connect with the other core. Others will not connect with either the other core or its probe. But some of them will connect with the probe of the other core. These probe-probe connections lead to gravity in the following way.

Such connections occur in two clusters, one centered around each of the cores. When such a connection occurs, it causes a force to be exerted on each core, pulling it toward the location of the connection.

The connections in one cluster can be almost anywhere around the core, except for a tiny cone extending from the core out in a direction opposite to the other core. The reason no connection can occur in that cone is quite simple: Take core 2 in Figure 6-1 below, for example. To get to a connection within the cone on the right side of particle 2, the probe from core 1 would have to pass close enough to core 2 to form a probe-core connection. This would prevent core 1's probe from forming any other connections on that cycle.

**Figure 6-1**
**Connection-free Zone of a Particle**

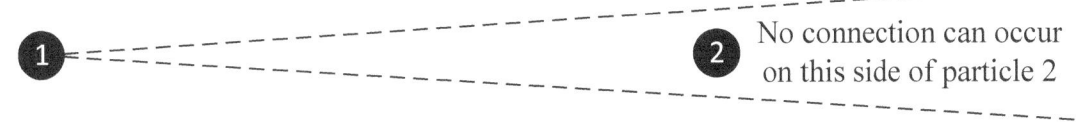

The zone to the right of core 2 will be called the nocon zone of particle 2.

Thus, each particle has a nocon zone. Probe-probe connections can occur anywhere around a particle except within its nocon zone.

Using the above figure as an example, consider a vertical plane through particle 2. Some probe-probe connections in the cluster about particle 2 will lie on the same side of that plane as particle 1. Each such connection will cause particle 2 to experience a slight attraction to particle 1, as indicated in Figure 6-2.

**Figure 6-2**
**When a Connection Occurs to the Left of Particle 2**

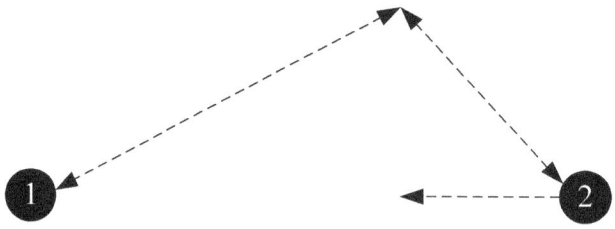

However, other probe-probe connections in that cluster will lie on the other side of that plane, and will cause particle 2 to experience a slight repulsion from particle 1, as indicated in Figure 6-3.

**Figure 6-3**
**When a Connection Occurs to the Right of Particle 2**

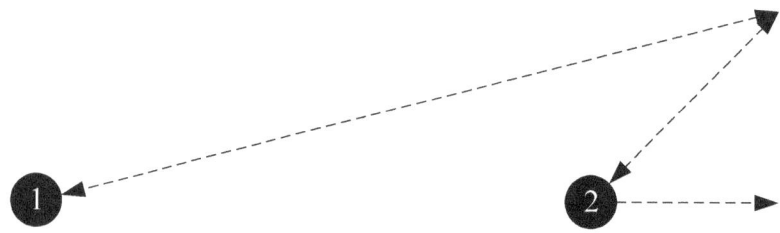

If nocon zones did not exist, then the number of probe-probe connections leading to attractive force on particle 2 would be the same as those leading to repulsive force on particle 2, and the net result would be zero force on particle 2. Thus gravity would not exist! However, the existence of the nocon zone of particle 2 means that there will be slightly more connections leading to attractive force on particle 2 than those leading to repulsive force.

And this slight discrepancy results in gravity!

Note: The forces involved in probe-probe connections are far larger than gravity. Gravity is an average of gazillions of probe-probe connections, which do not quite average out to zero. Hence there is no single interaction between particles which results in a force equal to gravity.

**Limitations on Gravity**

Connection theory recognizes several limitations on gravity, many of which are mentioned below.

On a single cycle, a probe may participate in one or more probe-probe connections, but at most only a single probe-core connection. After a probe-core connection, the probe cannot participate in any more connections on that cycle.

A gravitational (probe-probe) connection may occur anywhere on the path of a probe through another body. Such connections are extraordinarily rare, as can be shown by calculation. This is of course equivalent to the fact that the measured force of gravity is a very tiny fraction of electrical force.

Probes can only go so far from their cores. As discussed previously, the maximum distance that a probe can go is a bit less than 20 billion light years. Consequently, *no radiation can travel more than that distance*. Moreover, two astronomical bodies separated by 40 billion light years could have no effect whatsoever (gravitational or electromagnetic) on each other. (Why 40? Well, if the separation was only 39 billion light years, then conceivably a particle in one body could send out a probe toward the other body, and that other body could send out a probe toward the first body, and these probes could meet when each is 19.5 billion light years from its source.) Of course, a *particle* could travel from one body to the other body. Cosmic rays contain some particles which travel at near light speeds. Such particles could travel more than 20 billion light years.

Another limitation concerns the passage of a probe through dense matter. When a particle is in a star, its probe has an enormous number of possible cores and probes to connect with inside the star itself. It has been estimated that for a star such as our Sun, a probe from a particle on one side of the star which is aimed at the other side of the star is certain to make a probe-core connection before it reaches the other side of the star. Such a probe cannot make a gravitational connection with an object outside the star on that other side, since it never reaches outside the star.

Indeed, our Sun has a number of particles which cannot have any gravitational connection to a planet on the other side of the Sun from them. These particles have a total mass which has been estimated to be between three times the mass of Earth, and the mass of the planet Uranus. As a result of this, the apparent center of mass of the Sun is offset from the Sun's true center of mass by about 4,375 meters. This *gravitational offset* is the cause of the precession of the perihelia of the planets, as will be discussed on the next page.

Furthermore, the measured mass of the Sun is not the total mass of the Sun; but only its apparent mass.

This limitation also applies to very massive stars. Such a massive star cannot exert the full force of its gravity on other celestial bodies, since a substantial part of its gravity is absorbed by other particles within the star. In particular, a black hole cannot exert the full force of its gravity on other bodies.

Another limitation concerns Newton's hypothesis that every particle in the universe has a continuous gravitational attraction for every other particle. In connection theory, gravitational attraction between particles occurs by means of random and isolated events. Of course, in the long run, and within somewhat limited distances, the results are pretty much equivalent to Newton's idea, but the mechanism is quite different.

One more limitation involves rapidly rotating stars. Apart from such stars, the directions of probes are essentially random. However, for a rapidly rotating star such as a pulsar, more radiation occurs near the poles than elsewhere. This implies that the probes within the star have a preference for polar directions over equatorial directions. Consequently, objects in the equatorial plane of the star experience less gravitational attraction from the star than do objects above the poles of the star. Thus the apparent mass of such a star, estimated by gravitational measurements, would depend on the direction from which it is measured.

### A Correction to Newton's Law of Gravity

Newton's Law of gravity is usually given in terms of the force $F$ between two massive objects:

$$F = \frac{Gm_1 m_2}{r^2},$$

where $G$ is Newton's gravitational constant, $m_1$ and $m_2$ are the masses of the two objects, and $r$ is the distance between their centers of mass.

In view of the limitations mentioned above, these definitions must be interpreted as follows:

> $m_1$ and $m_2$ are the *apparent* masses of the two objects
>
> $r$ is the distance between their centers of mass, minus any appropriate gravitational offsets.

Also, the force $F$ must be applied at the apparent centers of mass of the two bodies, not at their true centers of mass. With these revised definitions, Newton's formula is exceptionally accurate.

### A Further Limitation on Gravity

Newton's law of gravity applies to massive bodies such as space craft and stars. However, it is sometimes incorrectly applied to particles, with the invalid result that the gravitational attraction between them goes to infinity as the distance between them goes to zero. But as mentioned on page 12, the gravitational attraction between them levels off to a finite value when the distance between them reaches a certain minimal amount or less.

### Perihelia Precessions

This part shows that the planets and asteroids that orbit the Sun sense an apparent center of mass in the Sun which is offset from its true center of mass. This difference is called the *gravitational offset* of the Sun, and is the cause of the precessions of the perihelia of the objects that orbit the Sun.

In 1858, Urbain Le Verrier discovered that the observed perihelion of Mercury's orbit had precessed; that is, its location differed by a small but significant amount from that predicted by Newton's law of gravity.

Le Verrier did not realize that this difference was caused not by a failure of Newton's law, but instead by an incorrect application of that law. As will be seen below, the correct application of that law yields the correct orbits of objects orbiting the Sun, including correct predictions of the precession of the perihelia of those objects.

In 1916, Einstein showed that his theory of general relativity was able to predict the amount of the precession. Einstein attributed the success of this prediction to the presumption that spacetime is curved, among other things.

Looking at the problem afresh, it would seem that Mercury and the other planets and asteroids that orbit the Sun behave as if they each sense an apparent center of mass in the Sun which is offset from its true center of mass (the *gravitational offset*). This point of view is suggested by the predictions of general relativity (GR) on the precessions of the perihelia of several planets and one asteroid. These precessions are shown in Table 6-1. Precessions are given in units of arc seconds per Julian century, which will be abbreviated as $\left(\text{arc sec}\left(\text{J-cent}\right)^{-1}\right)$ in the tables.

**Table 6-1**
**Perihelion Precessions Calculated by General Relativity**

| Planet or Asteroid | GR-Calculated Precession $\left(\text{arc sec}\left(\text{J-cent}\right)^{-1}\right)$ |
|---|---|
| Mercury | 42.98 |
| Venus | 8.62 |
| Earth | 3.84 |
| Mars | 1.35 |
| 1566 Icarus | 10.05 |
| Jupiter | 0.0623 |
| Saturn | 0.0137 |
| Neptune | 0.0008 |

There is a relation between these precession values and the corresponding gravitational offsets. To see this, first consider the sketch shown in Figure 6-4. Of course, this sketch is not to scale.

**Figure 6-4**
**An Illustration of a Gravitational Offset**

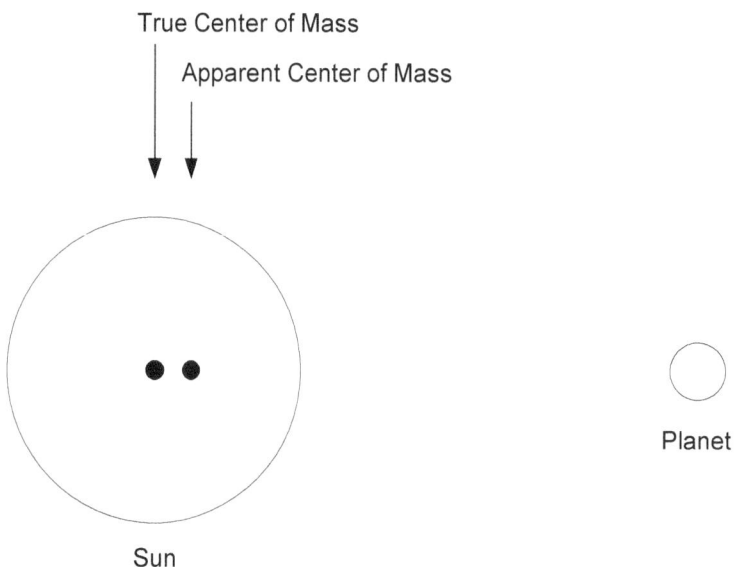

The gravitational offset is the distance between the true center of mass of the Sun and its apparent center of mass as sensed by the planet. The variable $q$ will be used to represent this distance.

Classical celestial mechanics in ordinary Euclidean space can be used to derive the orbit which results from the existence of a gravitational offset. This leads to the following equation, which relates the precession of the perihelion per orbit to the gravitational offset $q$. The derivation of this equation is found in Chapter 7.

(6-1) $\qquad \text{Precession per orbit} = 360 \times 3600 \left( \dfrac{1}{\sqrt{1-2Lq}} - 1 \right)$ seconds of arc.

In this equation, the parameter $L$ is defined as

(6-2) $\qquad L = \dfrac{G(M_1 + M_2)T^2}{4\pi^2 a^4 (1-e^2)},$

where

$\qquad G \qquad$ is Newton's gravitational constant,

$\qquad M_1 \qquad$ is the apparent mass of the Sun,

$M_2$     is the mass of the planet,
$T$      is the time of one orbit,
$a$      is the length of the semimajor axis of the orbit,
$e$      is the eccentricity of the orbit.

Results from equation (6-1) can easily be transformed into units of arc seconds per Julian century.

Equation (6-1) can be used to find the gravitational offsets corresponding to the numbers in Table 6-1. The results are given in Table 6-2.

**Table 6-2**
**Planetary Precessions and Corresponding Gravitational Offsets**

| Planet or Asteroid | GR-Calculated Precession $\left(\text{arc sec}\left(\text{J-cent}\right)^{-1}\right)$ | Gravitational Offset (m) |
|---|---|---|
| Mercury | 42.98 | 4,430.066 |
| Venus | 8.62 | 4,427.607 |
| Earth | 3.84 | 4,431.665 |
| Mars | 1.35 | 4,427.129 |
| 1566 Icarus | 10.05 | 4,427.124 |
| Jupiter | 0.0623 | 4,432.832 |
| Saturn | 0.0137 | 4,506.930 |
| Neptune | 0.0008 | 4,601.689 |

These offsets are each about 6 millionths of the solar radius of the Sun. Thus they are relatively small, but yield a significant effect.

The close numerical agreement of the numbers in the third column is likely due to the fact that they are the results of the solution of a partial differential equation (general relativity). Even so, that agreement suggests that the reason for these offsets has nothing to do with the planets, but instead is a property of the Sun itself. (The agreement between the values in the last two rows with those in the upper rows is close, but not as good as that between the upper values themselves. Presumably this implies that these last two values are not as accurate as the ones above.)

If indeed this is a property of the Sun, then a single value should do for all the planets and asteroids of the solar system. An estimate of such a single value was obtained by taking the least-square average of the numbers in column 3; that is, a value $X$ which minimizes the sum of squares of the differences between the values in column 3 and $X$. The result is $X = 4,429.839$ meters.

Thus *general relativity calculations imply that the Sun has a gravitational offset of 4,429.839 meters, which is responsible for the precessions of the perihelia of planets and asteroids that orbit the Sun.*

Using this value in equation (6-1) gives the following predictions for the precessions of the perihelia for the objects in Table 6-1:

**Table 6-3**
**GR-Calculated versus Predicted Planetary Perihelion Precessions**

| Planet or Asteroid | GR-Calculated Precession $\left(\text{arc sec}\left(\text{J-cent}\right)^{-1}\right)$ | Predicted Precession $\left(\text{arc sec}\left(\text{J-cent}\right)^{-1}\right)$ |
|---|---|---|
| Mercury | 42.98 | 42.97779750 |
| Venus | 8.62 | 8.62434531 |
| Earth | 3.84 | 3.83841803 |
| Mars | 1.35 | 1.35082636 |
| 1566 Icarus | 10.05 | 10.05616332 |
| Jupiter | 0.0623 | 0.06225794 |
| Saturn | 0.0137 | 0.01346566 |
| Neptune | 0.0008 | 0.00077012 |

The agreement between columns 2 and 3 seems to substantiate the validity of the calculated values in column 2.

However, the *observed* value of the planet Mercury's perihelion precession is 42.4446 seconds of arc per Julian century. This value corresponds to a gravitational offset of 4,374.881 meters. Since the gravitational offset determines the Sun's contribution to the precession of the perihelion for every planet and asteroid which orbits the Sun, it is reasonable to see what precessions are implied by this gravitational offset. The results are shown in Table 6-4.

**Table 6-4**
**Perihelion Precessions Corresponding to**
**the Observed Gravitational Offset of Mercury**

| Planet or Asteroid | Calculated Precession $\left(\text{arc sec}\left(\text{J-cent}\right)^{-1}\right)$ |
|---|---|
| Mercury | 42.4446 |
| Venus | 8.517349 |
| Earth | 3.790797 |
| Mars | 1.334068 |
| 1566 Icarus | 9.931404 |
| Jupiter | 0.061486 |
| Saturn | 0.013299 |
| Neptune | 0.000761 |

These values are each about 1.2% smaller than those derived from general relativity. The value for Mercury is the observed value. I have not come across any observed values for the other bodies in the table. If there are any such observations available, it would be very interesting to see how they compare to these values!

Thus it is clear that a gravitational offset exists in the Sun, and that it is the cause of the precessions of the perihelia of the planets and asteroids that orbit the Sun. There are two different estimates of this gravitational offset, one from general relativity, and one based on a single observation. Of course, that observation is of the most prominent precession (of Mercury), and should carry some weight.

At any rate, the exact value of the Sun's gravitational offset determines the precessions of the perihelia of the objects that orbit the Sun, and is also important for making highly precise calculations of the Sun's gravitational influence on such things as spacecraft.

Thus there are two models which can be used to explain the precession of the perihelia of the objects which orbit the Sun: general relativity, and gravitational offsets.

But whether or not the gravitational offset model is the correct one, it seems inescapable that a gravitational offset exists in the Sun.

### Gravity in the Solar System

The Sun is the only object in the solar system which has a gravitational offset. However, the planets themselves actually absorb some of the Sun's gravity. Thus when the moon of a planet lies in the planet's shadow, it will experience less gravity from the Sun than when not in the shadow.

So calculations of gravitational effects in the solar system fall into three categories. The first is calculations involving two bodies neither of which is the Sun. In this case, classical calculations using Newton's law of gravity yield correct results.

The second category is calculations between the Sun and another body. Here, Newton's law must be applied as interpreted in equation (6-4).

The third category is calculations of the orbit of a moon around its planet. In this case, it is appropriate to take into account the effect of the lessening of Sun's gravity on the moon as it passes through the shadow of its planet. At present, there is no information available on the amount of this lessening.

### Why?

Having concluded that a gravitational offset exists in the Sun, the next question is: What causes it?

The mere existence of the gravitational offset implies that there are present in the Sun a substantial number of particles that do not exert any gravitational force on a planet which is on the opposite side of the Sun. More generally, for any planet or asteroid, call it Y, there are a substantial number of particles in the Sun that do not exert any gravitational force on Y, and which lie on the side of the Sun opposite to Y.

The obvious conclusion is that the mass of the remaining particles in the Sun blocks the gravitational force of these particles from emerging from the Sun.

To explain this, consider the connection theory model of a primitive particle. It emits a probe which is ready to engage in connections. If the particle is in a star, then the probe has countless numbers of probes and cores with which to connect. Evidence from the Sun indicates that when a primitive particle on one side of the Sun emits a probe toward the opposite side of the Sun, that probe has a 100% chance of making a probe-core connection before it reaches the other side of the Sun. Thus it is unable to engage in any more connections on that cycle, and in particular is unable to make any gravitational (probe-probe) connections outside the Sun on that cycle.

## CHAPTER 7:  DERIVATION OF EQUATION (6-1)

### Differential Equations of Motion

The key idea which explains the shift in the perihelia of planets is this:  When two bodies are in orbit around each other, two of the things which determine their orbits are not their centers of mass, but instead their *apparent* centers of mass.  They sense each other's presence gravitationally, but each senses the center of mass of the other at a point possibly offset from its true center of mass.  These offsets determine their mutual orbits.

Suppose two objects of gravitational masses $M_1$ and $M_2$ move in orbits around each other.  Denote the positions of their centers of mass by the vectors $\mathbf{P}_1(t)$ and $\mathbf{P}_2(t)$.  Let $\mathbf{P} = \mathbf{P}_2 - \mathbf{P}_1$, let $p = |\mathbf{P}|$, and let $q_1$ and $q_2$ be the sizes of the gravitational offsets of the two objects.  Let $\mathbf{S}_1$ be the point between $\mathbf{P}_1(t)$ and $\mathbf{P}_2(t)$ which is a distance of $q_1$ from $\mathbf{P}_1(t)$, and let $\mathbf{S}_2$ be the point between $\mathbf{P}_1(t)$ and $\mathbf{P}_2(t)$ which is a distance of $q_2$ from $\mathbf{P}_2(t)$.

These assumptions are made:

**A7.1**   Each of the two objects acts as a rigid body.

**A7.2**   $\mathbf{P}_1$, $\mathbf{S}_1$, $\mathbf{S}_2$, and $\mathbf{P}_2$ lie on a straight line in that order.

**A7.3**   The offset distances $q_i$ do not change with time.

Let $q = q_1 + q_2$.

The gravitational attraction between the two objects is applied at the two points $\mathbf{S}_1$ and $\mathbf{S}_2$.  By the above assumptions, the distance between them is equal to $p - q$.  Let $\mathbf{F}_i$ denote the gravitational force on object $i$; then (by analogy with Newton's law)

$$(7\text{-}1) \qquad \mathbf{F}_1 = \frac{GM_1M_2}{(p-q)^2} \frac{\mathbf{P}}{p}$$

and $\mathbf{F}_2 = -\mathbf{F}_1$.

By the first and second assumptions, the forces act on the gravitational masses, and the force vectors are applied at the apparent centers of mass.  (One result is that the force on the apparent center of mass of the Sun contributes to the rotation of the Sun.)  Hence, the differential equations of motion are

(7-2) $\qquad M_i \mathbf{P}_i'' = \mathbf{F}_i$,

where primes denote time derivatives. Since $\mathbf{F}_2 = -\mathbf{F}_1$,

$$M_1 \mathbf{P}_1'' + M_2 \mathbf{P}_2'' = 0,$$

and so the center of mass of the entire system:

$$\frac{M_1 \mathbf{P}_1 + M_2 \mathbf{P}_2}{M_1 + M_2}$$

moves in a straight line.

Now consider the differential equation for $\mathbf{P}$. Since $\mathbf{P} = \mathbf{P}_2 - \mathbf{P}_1$, it follows from equation (7-1) and (7-2) that

(7-3) $\qquad \mathbf{P}'' = \mathbf{P}_2'' - \mathbf{P}_1'' = \dfrac{\mathbf{F}_2}{M_2} - \dfrac{\mathbf{F}_1}{M_1} = \dfrac{-K\mathbf{P}}{p(p-q)^2}$,

where

$$K = G(M_1 + M_2).$$

Taking the vector product of $\mathbf{P}$ with the first and last sides of equation (7-3), it is found that $\mathbf{P} \times \mathbf{P}'' = 0$, which can be integrated directly to yield the fact that $\mathbf{P} \times \mathbf{P}'$ equals a constant vector. Thus, the orbits lie in a plane perpendicular to that vector. Take that plane as the $x$-$y$ plane, and let the origin be at $\mathbf{P}_1$.

### *Vis Viva* Integral

Put equation (7-3) in rectangular coordinates:

$$x'' = \frac{-Kx}{p(p-q)^2},$$

$$y'' = \frac{-Ky}{p(p-q)^2}.$$

Multiply the first equation by $2x'$, the second by $2y'$, and add:

(7-4)        $2x'x'' + 2y'y'' = \dfrac{-2K(xx'+yy')}{p(p-q)^2}$ .

But by definition, the square of the velocity $v$ is given by

$$v^2 = x'^2 + y'^2 ,$$

and so the left side of (7-4) equals $(v^2)'$ . Also, $p = \sqrt{x^2+y^2}$ and $q$ is a constant. Hence,

$$(p-q)' = p' = \frac{2xx'+2yy'}{2\sqrt{x^2+y^2}} = \frac{xx'+yy'}{p} ,$$

and so the right side of (7-4) is equal to

$$\frac{-2K(p-q)'}{(p-q)^2} .$$

Thus, equation (7-4) can be integrated to obtain

$$v^2 = \frac{2K}{p-q} + c_1$$

for some constant $c_1$ .

### The Main Differential Equation

Using polar coordinates, set

$$\mathbf{P} = p(\cos\theta, \sin\theta) .$$

Then the second derivative is

$$\mathbf{P}'' = (p'' - p\theta'^2)(\cos\theta, \sin\theta)$$
$$+ (2p'\theta' + p\theta'')(-\sin\theta, \cos\theta).$$

and the differential equation (7-3) becomes

$$\left( p'' - p\theta'^2 + \frac{K}{(p-q)^2} \right)(\cos\theta, \sin\theta)$$
$$+ (2p'\theta' + p\theta'')(-\sin\theta, \cos\theta) = 0.$$

Since $(\cos\theta, \sin\theta)$ and $(-\sin\theta, \cos\theta)$ are mutually perpendicular unit vectors, the above sum can vanish only if the coefficients of those vectors are each zero; thus

(7-5)          $$p'' - p\theta'^2 + \frac{K}{(p-q)^2} = 0$$

and

$$2p'\theta' + p\theta'' = 0.$$

From the latter equation, it follows that

$$(p^2\theta')' = p(2p'\theta' + p\theta'') = 0,$$

and so for some constant $h$,

(7-6)          $$p^2\theta' = h.$$

Using this fact in equation (7-5) yields

(7-7)          $$p'' - \frac{h^2}{p^3} + \frac{K}{(p-q)^2} = 0.$$

Next, use the standard transformation $p = 1/u$, and let the angle $\theta$ replace time as the independent variable, with dots indicating derivatives with respect to $\theta$. By the definition of $u$, along with equation (7-6), it follows that

$$p' = \frac{-u'}{u^2} = \frac{-\dot{u}\theta'}{u^2} = -\dot{u}p^2\theta' = -h\dot{u}.$$

One more use of equation (7-6) yields

$$p'' = -h\ddot{u}\theta' = -h^2u^2\ddot{u}.$$

Hence, rewriting equation (7-7) in terms of $u$ rather than $p$ and dividing the result by $-h^2u^2$ yields

(7-8)          $$\ddot{u} + u = \frac{L}{(1-qu)^2},$$

where

$$L = \frac{K}{h^2}.$$

### Approximation 1

Equation (7-6) represents twice the rate of accumulation of area in polar coordinates. The integral of (7-6) over one orbit (say for $t = 0$ to $T$, where $T$ is the time required for the orbit) yields twice the area contained within the orbit. Under the (usually very accurate) approximation that the orbit is an ellipse, that area equals $\pi ab$, where $a$ and $b$ are the lengths of the semimajor and semiminor axes. Hence, the integral results in the equation

$$2\pi ab = Th,$$

so that

$$h = \frac{2\pi ab}{T} = \frac{2\pi a^2 \sqrt{1 - e^2}}{T},$$

where $e$ is the eccentricity of the ellipse. Consequently,

$$L = \frac{K}{h^2} = \frac{G(M_1 + M_2)T^2}{4\pi^2 a^4 (1 - e^2)}.$$

### Approximation 2

Each offset $q_i$ is a fraction of the radius of the corresponding star or planet, and so the sum of the offsets, $q$, is much smaller than the separation $p$ between the two objects. Thus the quantity $q/p = qu$ is much smaller than 1, and so for all $u$ of interest, the denominator on the right side of equation (7-8) can be expanded into a power series which converges rapidly. In fact, an excellent approximation is obtained by dropping all terms of order $u^2$ and higher. The result is the approximate equation

$$\ddot{u} + u \cong L(1 + 2qu),$$

or

$$\ddot{u} + (1 - 2Lq)u \cong L.$$

The nature of the solution to this equation depends on whether $2Lq$ is smaller or larger than 1. It is not correct to say that $2Lq$ is small merely because $qu$ is small. However, for the solution to be at all meaningful, $2Lq$ must be less than 1; otherwise, the solution would not be periodic, but would either expand or contract exponentially, which is not of interest.  Hence for the moment, assume that $2Lq$ is less than 1; it is easy to show that it is positive.  This leads to the solution

$$u = A\cos\left(\sqrt{1-2Lq}\,\theta - \omega\right) + \frac{L}{1-2Lq}$$

for constants $A$ and $\omega$, as is easily verified by differentiation.  From this, it is possible to solve for $p$:

(7-9) $$p = \frac{1}{A\cos\left(\sqrt{1-2Lq}\,\theta - \omega\right) + \dfrac{L}{1-2Lq}} .$$

For small values of $2Lq$, this equation is very nearly that of an ellipse.

Recall that $p$ represents the distance of the true center mass of the second object from that of the first object in a coordinate system in which the origin is at the true center of mass of the first object. The minimum value of $p$ occurs when the denominator in equation (7-9) reaches its maximum, and that happens when the cosine equals 1.  Similarly, the maximum value of $p$ occurs when the denominator reaches its minimum, and that happens when the cosine equals $-1$.  Of course, this presumes that the denominator never vanishes for any value of $\theta$.  It will be seen below that this is the case.

For an ellipse, the separation $p$ at the time of periastron is equal to $a(1-e)$, and at apastron, $p$ equals $a(1+e)$.  Using the fact that equation (7-9) is very nearly that of an ellipse, form the equations

$$A + \frac{L}{1-2Lq} = \frac{1}{a(1-e)},$$

$$-A + \frac{L}{1-2Lq} = \frac{1}{a(1+e)}.$$

These can be combined to yield

$$\frac{L}{1-2Lq} = \frac{1}{a(1-e^2)}$$

and

$$A = \frac{e}{a(1-e^2)} .$$

The interval between two periastra is found by taking two successive values of $\theta$ for which the argument of the cosine is equal to a multiple of $2\pi$. Two such values are $\theta_1$ and $\theta_2$, where $\theta_1 < \theta_2$, and where

$$\sqrt{1-2Lq}\,\theta_1 - \omega = 0$$

and

$$\sqrt{1-2Lq}\,\theta_2 - \omega = 2\pi .$$

for some value $\omega$.

The change in $\theta$ from one periastron to the next is thus

$$\theta_2 - \theta_1 = \frac{2\pi + \omega}{\sqrt{1-2Lq}} - \frac{\omega}{\sqrt{1-2Lq}}$$

$$= \frac{2\pi}{\sqrt{1-2Lq}}.$$

If this change were equal to $2\pi$, there would be no shift in periastron. Otherwise, the shift in periastron is equal to the above amount minus $2\pi$:

$$2\pi\left(\frac{1}{\sqrt{1-2Lq}} - 1\right) \qquad \text{in radians}$$

$$360\left(\frac{1}{\sqrt{1-2Lq}} - 1\right) \qquad \text{in degrees}$$

(7-10) $\qquad 360\times 3600\left(\frac{1}{\sqrt{1-2Lq}} - 1\right) \qquad \text{in arc seconds.}$

Thus equation (6-1), which is the same as equation (7-10) above, is established.